# Plasmon Assisted Near-Field
Manipulation and Photocatalysis

# Plasmon Assisted Near-Field Manipulation and Photocatalysis

Editor

**Zhenglong Zhang**

Basel • Beijing • Wuhan • Barcelona • Belgrade • Novi Sad • Cluj • Manchester

*Editor*
Zhenglong Zhang
Shaanxi Normal University
Xi'an
China

*Editorial Office*
MDPI
St. Alban-Anlage 66
4052 Basel, Switzerland

This is a reprint of articles from the Special Issue published online in the open access journal *Nanomaterials* (ISSN 2079-4991) (available at: https://www.mdpi.com/journal/nanomaterials/special_issues/plasmon_assisted_catalysis).

For citation purposes, cite each article independently as indicated on the article page online and as indicated below:

Lastname, A.A.; Lastname, B.B. Article Title. *Journal Name* **Year**, *Volume Number*, Page Range.

**ISBN 978-3-0365-8320-4 (Hbk)**
**ISBN 978-3-0365-8321-1 (PDF)**
**doi.org/10.3390/books978-3-0365-8321-1**

Cover image courtesy of Zhenglong Zhang

© 2023 by the authors. Articles in this book are Open Access and distributed under the Creative Commons Attribution (CC BY) license. The book as a whole is distributed by MDPI under the terms and conditions of the Creative Commons Attribution-NonCommercial-NoDerivs (CC BY-NC-ND) license.

# Contents

**Zhenglong Zhang**
Editorial for Special Issue "Plasmon Assisted Near-Field Manipulation and Photocatalysis"
Reprinted from: *Nanomaterials* **2023**, *13*, 1427, doi:10.3390/nano13081427 . . . . . . . . . . . . . . . 1

**Xiaohua Wang, Chengyun Zhang, Xilin Zhou, Zhengkun Fu, Lei Yan, Jinping Li, et al.**
Plasmonic Effect of Ag/Au Composite Structures on the Material Transition
Reprinted from: *Nanomaterials* **2022**, *12*, 2927, doi:10.3390/nano12172927 . . . . . . . . . . . . . . . 5

**Xilin Zhou, Huan Chen, Baobao Zhang, Chengyun Zhang, Min Zhang, Lei Xi, et al.**
Plasmon Driven Nanocrystal Transformation by Aluminum Nano-Islands with an Alumina Layer
Reprinted from: *Nanomaterials* **2023**, *13*, 907, doi:10.3390/nano13050907 . . . . . . . . . . . . . . . 13

**Shali Lin, Xiaohu Mi, Lei Xi, Jinping Li, Lei Yan, Zhengkun Fu, et al.**
Efficient Reduction Photocatalyst of 4-Nitrophenol Based on Ag-Nanoparticles-Doped Porous ZnO Heterostructure
Reprinted from: *Nanomaterials* **2022**, *12*, 2863, doi:10.3390/nano12162863 . . . . . . . . . . . . . . . 23

**Quanjiang Li, Jingang Wang, Shenghui Chen and Meishan Wang**
Impurity Controlled near Infrared Surface Plasmonic in AlN
Reprinted from: *Nanomaterials* **2022**, *12*, 459, doi:10.3390/nano12030459 . . . . . . . . . . . . . . . 33

**Yunyan Wang, Chen Zhou, Yiping Huo, Pengfei Cui, Meina Song, Tong Liu, et al.**
Efficient Excitation and Tuning of Multi-Fano Resonances with High Q-Factor in All-Dielectric Metasurfaces
Reprinted from: *Nanomaterials* **2022**, *12*, 2292, doi:10.3390/nano12132292 . . . . . . . . . . . . . . . 45

**Shaobo Ge, Weiguo Liu, Xueping Sun, Jin Zhang, Pengfei Yang, Yingxue Xi, et al.**
Efficient Achromatic Broadband Focusing and Polarization Manipulation of a Novel Designed Multifunctional Metasurface Zone Plate
Reprinted from: *Nanomaterials* **2021**, *11*, 3436, doi:10.3390/nano11123436 . . . . . . . . . . . . . . . 59

**Jihua Xu, Jinmeng Li, Guangxu Guo, Xiaofei Zhao, Zhen Li, Shicai Xu, et al.**
Facilely Flexible Imprinted Hemispherical Cavity Array for Effective Plasmonic Coupling as SERS Substrate
Reprinted from: *Nanomaterials* **2021**, *11*, 3196, doi:10.3390/nano11123196 . . . . . . . . . . . . . . . 71

**Daobin Luo, Pengcheng Hong, Chao Wu, Shengbo Wu and Xiaojing Liu**
Optical Properties of Ag Nanoparticle Arrays: Near-Field Enhancement and Photo-Thermal Temperature Distribution
Reprinted from: *Nanomaterials* **2022**, *12*, 3924, doi:10.3390/nano12213924 . . . . . . . . . . . . . . . 85

**Yi Cao, Jing Li, Mengtao Sun, Haiyan Liu and Lixin Xia**
Nonlinear Optical Microscopy and Plasmon Enhancement
Reprinted from: *Nanomaterials* **2022**, *12*, 1273, doi:10.3390/nano12081273 . . . . . . . . . . . . . . . 99

**Chengyun Zhang, Jianxia Qi, Yangyang Li, Qingyan Han, Wei Gao, Yongkai Wang, et al.**
Surface-Plasmon-Assisted Growth, Reshaping and Transformation of Nanomaterials
Reprinted from: *Nanomaterials* **2022**, *12*, 1329, doi:10.3390/nano12081329 . . . . . . . . . . . . . . . 119

Editorial

# Editorial for Special Issue "Plasmon Assisted Near-Field Manipulation and Photocatalysis"

Zhenglong Zhang

School of Physics and Information Technology, Shaanxi Normal University, Xi'an 710062, China; zlzhang@snnu.edu.cn

Citation: Zhang, Z. Editorial for Special Issue "Plasmon Assisted Near-Field Manipulation and Photocatalysis". *Nanomaterials* 2023, 13, 1427. https://doi.org/10.3390/nano13081427

Received: 11 April 2023
Accepted: 19 April 2023
Published: 21 April 2023

Copyright: © 2023 by the author. Licensee MDPI, Basel, Switzerland. This article is an open access article distributed under the terms and conditions of the Creative Commons Attribution (CC BY) license (https:// creativecommons.org/licenses/by/ 4.0/).

Accurately establishing the near field is crucial to enhancing optical manipulation and resolution, and is pivotal to the application of nanoparticles in the field of photocatalysis. A novel type of modulated optical field that enables the multi-dimensional control of the amplitude, polarization and phase can be obtained via the precise manipulation of the plasmonic near field of metal nanostructures. Meanwhile, the energy stored in the plasmonic field can induce hot carriers in the metal, which ultimately dissipate by coupling to the phonon modes of the metal nanoparticles, resulting in an elevated lattice temperature.

The plasmonic nearfield, hot carriers and their heating effects can catalyze the chemical reactions of reactants, including molecules and nanomaterials. Firstly, the chemical efficiency can be enhanced by the plasmonic near field due to its elevated photon density. Secondly, the hot carriers induced by plasmon decay can transfer to the reactant via the indirect electron transfer or direct electron excitation process, and the specific chemical reaction channels can be selectively enhanced by controlling the energy distribution of hot carriers. Thirdly, the local thermal effect that is followed by plasmon decay offers opportunities to facilitate the chemical reactions of molecules and induce the crystal growth and transformation of nanomaterials at room temperature. As a new class of photocatalysts, plasmonic noble metal nanoparticles that possess the unique ability to harvest light energy across the entire visible spectrum and produce effective energy conversion have been explored as a promising novel direction in the amelioration of the energy crisis. These opportunities have motivated this Special Issue, entitled 'Plasmon-Assisted Near-Field Manipulation and Photocatalysis', which has attracted research and review papers related to a variety of emerging nanomaterials, as outlined below.

Noble metal (Au/Ag/Cu/Al) nanostructures can produce surface plasmon resonance to promote or facilitate chemical reactions, as well as photocatalytic materials. In particular, Ag/Au nanoislands (NIs) and Ag NIs/Au film composite systems were designed, and their thermo-catalysis performance was investigated using the luminescence of $Eu^{3+}$ as a probe [1]. It was discovered that the metal NIs can also generate strong localized heat in low-temperature environments, enabling the transition of $NaYF_4$ to $Y_2O_3$. Furthermore, anti-oxidation was realized by depositing gold on the surface of silver, resulting in the relative stability of the constructed complex. These investigations can provide an enhanced understanding of the surface plasmon catalysis process and extend the potential applications of metal NIs. For utilization in an alternative application, self-assembled Al NIs with a thin alumina layer were designed with a plasmonic photothermal structure in order to achieve nanocrystal transformation via multi-wavelength excitation [2]. The Al NIs with an alumina layer demonstrated excellent photothermal conversion efficiency even in low-temperature environments, and their efficacy did decline significantly after storage in air for three months. Such an inexpensive $Al/Al_2O_3$ structure with a multi-wavelength response provides an efficient platform through which to achieve rapid nanocrystal transformation and fulfil its potential application in the wide-band absorption of solar energy.

Oxide-supported noble metal nanoparticles, which are one of the primary photocatalytic nanomaterials, have been investigated for their potential ability to enhance the stability and diminish the cost of photocatalysts. In particular, an Ag-nanoparticles-doped porous ZnO photocatalyst was prepared [3]. Under visible light irradiation, the heterostructure showed excellent catalytic activity over 4-nitrophenol due to the hot electrons induced by the localized surface plasmon resonance of Ag nanoparticles; this provides a novel heterostructure photocatalyst with the potential to be applied in solar energy and pollutant disposal. In addition, the optical properties of substitutional-doped aluminum nitride (AlN) were studied via multi-scale computational simulation methods, combined with density functional theory and finite element analysis [4]. It was discovered that a strong AlN surface plasmon resonance could be obtained in the near-infrared region by applying various alkali metal doping configurations, which not only improve the application of multi-scale computational simulations in quantum surface plasmons, but also promote the application of AlN in the field of surface-enhanced linear and non-linear optical spectroscopy.

Due to the large inherent loss of metals that occurs in phase matching, and its further limitations, a larger Q-factor cannot be obtained by utilizing traditional optical cavity modes and devices based on surface plasmon resonance. A silicon square-hole nano-disk array device was proposed in order to realize multi-Fano resonances with a high Q-factor, narrow line width, large modulation depth and enhanced near-field enhancement, which could provide the basis for the application of a novel method in the realms of multi-wavelength communications, lasing, and nonlinear optical devices [5]. Then, a nested composite structured multifunctional metasurface zone plate was designed and fabricated by integrating the metasurface onto the surface of the multi-level diffraction lens rings [6]. Based on the global optimization mathematical iterative method, the height distribution of the multifunctional metasurface zone plate was optimized in order to realize an extremely efficient achromatic broadband focus. This combination enhances the degree of freedom that exists in micro–nano optical design, and is expected to be applied in multifunctional focusing devices, polarization imaging, and various other fields. Lastly, the self-organizing process of component and array manufacturing was combined with imprinting technology in order to construct a cheap and reproducible flexible polyvinyl alcohol nanocavity array that is decorated with the silver nanoparticles [7]. The substrate exhibited excellent mechanical stability in bending experiments, and was able to achieve low-cost, high-sensitivity, uniform and favorable surface-enhanced Raman scattering detection, particularly in regard to situ detection; furthermore, it demonstrated promise in food safety and biomedicine applications.

As a plasmonic photocatalytic mechanism, the photothermal properties of nanomaterials have received widespread attention due to their broad applicative potential. The near-field and photo-thermal temperature distribution of a nanoparticle array was numerically investigated by considering the scattering light field among particles [8]. It has been determined that the position of the 'hot spots' does not rotate with the polarization direction of the incident light and always remains in the particle gaps along the line between the particle centers, which provides theoretical considerations for the near-field manipulation and photo-thermal applications of nanoarrays.

Finally, it was determined that the surface plasmon could strongly confine electromagnetic fields near the metal nanostructures in order to generate a localized near field, which has been widely utilized in surface-enhanced spectroscopy and nonlinear optics. An overview of the mechanism of surface plasmon and of near-field nonlinear effects is offered in one paper published in this volume, and describes some of the latest research that focuses on the applications of nonlinear optical microscopy systems [9]. Furthermore, the enhanced near field, hot carriers and localized thermal effect play an important role in photocatalysis. The other review paper is focused upon surface-plasmon-assisted photocatalysis, including nanomaterial reshaping, growth and transformation [10]. The current status of and perspectives on the future of plasmonic photocatalysis are reviewed, which will promote the development of surface plasmon in regard to the regulation of nanomaterials.

Overall, this volume provides a selected collection of papers that covers various aspects of plasmonic near-field manipulation and photocatalysis; we sincerely hope that the reader will benefit from such an informative and insightful Special Issue.

**Acknowledgments:** I am grateful to all the authors who contributed to this Special Issue. We also acknowledge the referees for reviewing the manuscripts.

**Conflicts of Interest:** The author declares no conflict of interest.

## References

1. Wang, X.; Zhang, C.; Zhou, X.; Fu, Z.; Yan, L.; Li, J.; Zhang, Z.; Zheng, H. Plasmonic Effect of Ag/Au Composite Structures on the Material Transition. *Nanomaterials* **2022**, *12*, 2927. [CrossRef] [PubMed]
2. Zhou, X.; Chen, H.; Zhang, B.; Zhang, C.; Zhang, M.; Xi, L.; Li, J.; Fu, Z.; Zheng, H. Plasmon Driven Nanocrystal Transformation by Aluminum Nano-Islands with an Alumina Layer. *Nanomaterials* **2023**, *13*, 907. [CrossRef] [PubMed]
3. Lin, S.; Mi, X.; Xi, L.; Li, J.; Yan, L.; Fu, Z.; Zheng, H. Efficient Reduction Photocatalyst of 4-Nitrophenol Based on Ag-Nanoparticles-Doped Porous ZnO Heterostructure. *Nanomaterials* **2022**, *12*, 2863. [CrossRef] [PubMed]
4. Li, Q.; Wang, J.; Chen, S.; Wang, M. Impurity Controlled near Infrared Surface Plasmonic in AlN. *Nanomaterials* **2022**, *12*, 459. [CrossRef] [PubMed]
5. Wang, Y.; Zhou, C.; Huo, Y.; Cui, P.; Song, M.; Liu, T.; Zhao, C.; Liao, Z.; Zhang, Z.; Xie, Y. Efficient Excitation and Tuning of Multi-Fano Resonances with High Q-Factor in All-Dielectric Metasurfaces. *Nanomaterials* **2022**, *12*, 2292. [CrossRef] [PubMed]
6. Ge, S.; Liu, W.; Sun, X.; Zhang, J.; Yang, P.; Xi, Y.; Zhou, S.; Zhu, Y.; Pu, X. Efficient Achromatic Broadband Focusing and Polarization Manipulation of a Novel Designed Multifunctional Metasurface Zone Plate. *Nanomaterials* **2021**, *11*, 3436. [CrossRef] [PubMed]
7. Xu, J.; Li, J.; Guo, G.; Zhao, X.; Li, Z.; Xu, S.; Li, C.; Man, B.; Yu, J.; Zhang, C. Facilely Flexible Imprinted Hemispherical Cavity Array for Effective Plasmonic Coupling as SERS Substrate. *Nanomaterials* **2021**, *11*, 3196. [CrossRef] [PubMed]
8. Luo, D.; Hong, P.; Wu, C.; Wu, S.; Liu, X. Optical Properties of Ag Nanoparticle Arrays: Near-Field Enhancement and Photo-Thermal Temperature Distribution. *Nanomaterials* **2022**, *12*, 3924. [CrossRef] [PubMed]
9. Cao, Y.; Li, J.; Sun, M.; Liu, H.; Xia, L. Nonlinear Optical Microscopy and Plasmon Enhancement. *Nanomaterials* **2022**, *12*, 1273. [CrossRef] [PubMed]
10. Zhang, C.; Qi, J.; Li, Y.; Han, Q.; Gao, W.; Wang, Y.; Dong, J. Surface-Plasmon-Assisted Growth, Reshaping and Transformation of Nanomaterials. *Nanomaterials* **2022**, *12*, 1329. [CrossRef] [PubMed]

**Disclaimer/Publisher's Note:** The statements, opinions and data contained in all publications are solely those of the individual author(s) and contributor(s) and not of MDPI and/or the editor(s). MDPI and/or the editor(s) disclaim responsibility for any injury to people or property resulting from any ideas, methods, instructions or products referred to in the content.

Article

# Plasmonic Effect of Ag/Au Composite Structures on the Material Transition

Xiaohua Wang [1], Chengyun Zhang [2], Xilin Zhou [1], Zhengkun Fu [1], Lei Yan [1], Jinping Li [1,*], Zhenglong Zhang [1] and Hairong Zheng [1,*]

[1] School of Physics and Information Technology, Shaanxi Normal University, Xi'an 710119, China
[2] School of Electronic Engineering, Xi'an University of Posts & Telecommunications, Xi'an 710121, China
* Correspondence: ljping@snnu.edu.cn (J.L.); hrzheng@snnu.edu.cn (H.Z.)

**Abstract:** Noble metal nanostructures can produce the surface plasmon resonance under appropriate photoexcitation, which can be used to promote or facilitate chemical reactions, as well as photocatalytic materials, due to their strong plasmon resonance in the visible light region. In the current work, Ag/Au nanoislands (NIs) and Ag NIs/Au film composite systems were designed, and their thermocatalysis performance was investigated using luminescence of $Eu^{3+}$ as a probe. Compared with Ag NIs, the catalytic efficiency and stability of surface plasmons of Ag/Au NIs and Ag NIs/Au film composite systems were greatly improved. It was found that the metal NIs can also generate strong localized heat at low temperature environment, enabling the transition of $NaYF_4:Eu^{3+}$ to $Y_2O_3:Eu^{3+}$, and anti-oxidation was realized by depositing gold on the surface of silver, resulting in the relative stability of the constructed complex.

**Keywords:** Ag/Au composite structures; surface plasmon resonance; photothermal effect; crystal transition

---

## 1. Introduction

Plasmon photocatalysis, as a new method to enhance the performance of semiconductor laser catalysis based on localized surface plasmon resonance (LSPR) effect, has attracted great attention in the past decade [1–5]. Compared with conventional thermally driven catalysis, plasmon catalysis can significantly reduce the reaction temperature and achieve the desired catalytic activity in a very short time. More importantly, photocatalysis also shows excellent stability and high selectivity under mild reaction conditions [6]. As a new family of photocatalysts, the catalytic performance of plasmon driven and enhanced photocatalytic and electrocatalytic reactions are highly dependent on the rational design of plasmon nanostructures.

The surface plasmon resonance (SPR) of metal nanostructures can be adjusted by selecting suitable plasmon materials [7,8], particles [9], composites [10], and morphologies [11], et al. As the most common and effective surface plasmon material, Au, Ag and Cu have strong optical adsorption capacity in the visible region [2,6,12]. Au is used for catalysis due to its unique stability and excellent catalytic properties, but it is expensive [13,14]. Ag has advantages with intense electromagnetic field enhancement from a larger extinction cross-section, along with a narrow plasmon linewidth; it is often used as catalyst [15], but the catalytic efficiency of Ag nanostructures gradually weakens with the time stored in air. Thus, the investigation on the plasmon composite structures and their properties that can enhance the advantages and reduce weaknesses of a single element is necessary.

For the study of bimetals, Taerin Chung et al. reported the transfer of metal nanoislands from glass to other different substrates using various dewetting methods, enabling high-throughput and low-cost control and applications of metal nanoislands on different substrates provides direction [16]. Kateryna Loza et al. also used chemical reduction and laser burning methods to obtain the alloy and studied the characterization of the alloy [17].

In these reports, the preparation methods of nanoislands are still relatively complicated, and the influencing factors of plasmonic photothermal effect of nanoislands need further study. Our previous work investigated the photothermal properties of surface plasmon polaritons (SPPs) on metallic NIs by Au [18,19] to induce rapid crystal transitions, which can monitor the local temperature of metal nanoparticles, so here we also use this method to study the plasmonic photothermal effect of bimetallic nanoislands.

In this work, using the method of thermal evaporation, we hope to construct a stable bimetallic nano-island structure, and choose to deposit Au on the Ag surface for the purpose of anti-oxidation. The surface plasmon catalytic effect of Ag NIs and bimetal structures that include the nanoislands formed by Ag/Au (Ag/Au NIs) and Ag NIs covered with Au film (Ag NIs/Au film) were investigated by monitoring the fluorescence of $Eu^{3+}$. It is found that the plasmonic photothermal effect of metal nanoislands can be controlled by the annealing temperature, ambient temperature and the size of the nanoislands. Additionally, compared with Ag NIs, Ag/Au NIs and Ag NIs/Au film present higher catalytic efficiency and better stability, and significant localized heat generated by LSPR of NIs is capable of driving crystal transitions even at low temperature environment. These investigations can provide a better understanding of the surface plasmon catalysis and extend possible applications of metal NIs.

## 2. Materials and Methods

Ag/Au NIs was prepared on a pre-cleaned glass substrate through high vacuum evaporation. The schematic in Figure 1 shows the preparation procedure of Ag/Au composite NIs structures, in which the annealing was performed in air at RT–400 °C for 30 s. The specific preparation steps were as follows: first, 15 nm Ag film was evaporated on the substrate. Ag NIs/Au film was obtained by annealing the Ag film to obtain Ag NIs, and then a layer of 12 nm Au film was deposited on the surface; then the same annealing treatment was performed. During the evaporation process, the vacuum degree of the vacuum coater was $2.4 \times 10^{-4}$ Pa, and the deposition rate was 0.03 Å/s, and the metal targets were Au wires and Ag nanoparticles with a purity of 99.999%. Polycrystalline $NaYF_4$:$Eu^{3+}$ particles were synthesized by wet chemical method. All reagents, including Ln $(NO_3)_3$ (Ln = Y, Eu) (99.9%) and NaF (98%), as well as solvents, were purchased from Sigma-Aldrich Chemicals Co. (Shanghai, China), and used without any further treatment. During the experimental study, the $NaYF_4$ particles were evenly spread on the metal film.

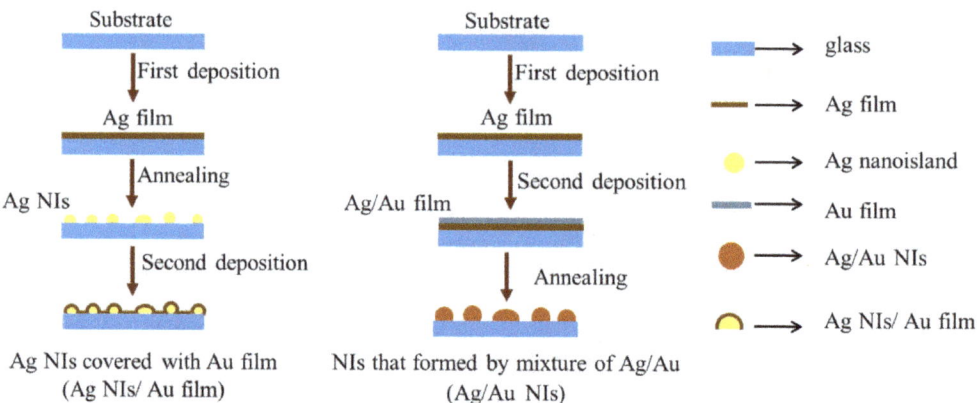

**Figure 1.** Preparation process of Ag/Au NIs and Ag NIs/Au film.

## 3. Results and Discussion

To investigate the thermocatalytic efficiency of metal NIs film structure, we prepared two types of composite nanosystems, Ag/Au NIs and Ag NIs/Au film, by different deposition and annealing processes. Ag NIs is prepared as the control group that is annealed, at 200 °C, for 30 s. As shown in Figure 2a–f, the AFM characterization results of Ag NIs, Ag/Au NIs and Ag NIs/Au film indicate that the averaged sizes of the island are 25 nm, 20 nm and 30 nm, respectively, and different colors under natural light are presented. In Figure 2g, compared with Ag NIs, the UV–Vis absorption spectra shows that Ag/Au composite nanostructures have a broadened spectral band width and a red-shifted spectral peak position, and the LSPR peak is located at 480 nm, which can better match the irradiation wavelength of 532 nm. The elemental composition of the Ag/Au composite system, Ag/Au NIs and Ag NIs/Au film were determined using a high Angle Ring dark field scanning transmission electron microscope (HAADF-STEM) and the results areas shown in Figure 2h,i. The energy dispersive X-ray (EDX) elemental mapping analysis further demonstrated the microstructure of the composite nanoislands. The characterization results of Figure 2 show that the Ag/Au composite systems have a diverse microstructure compared with Ag NIs, which further leads to its absorption spectrum being adjusted in a wide range.

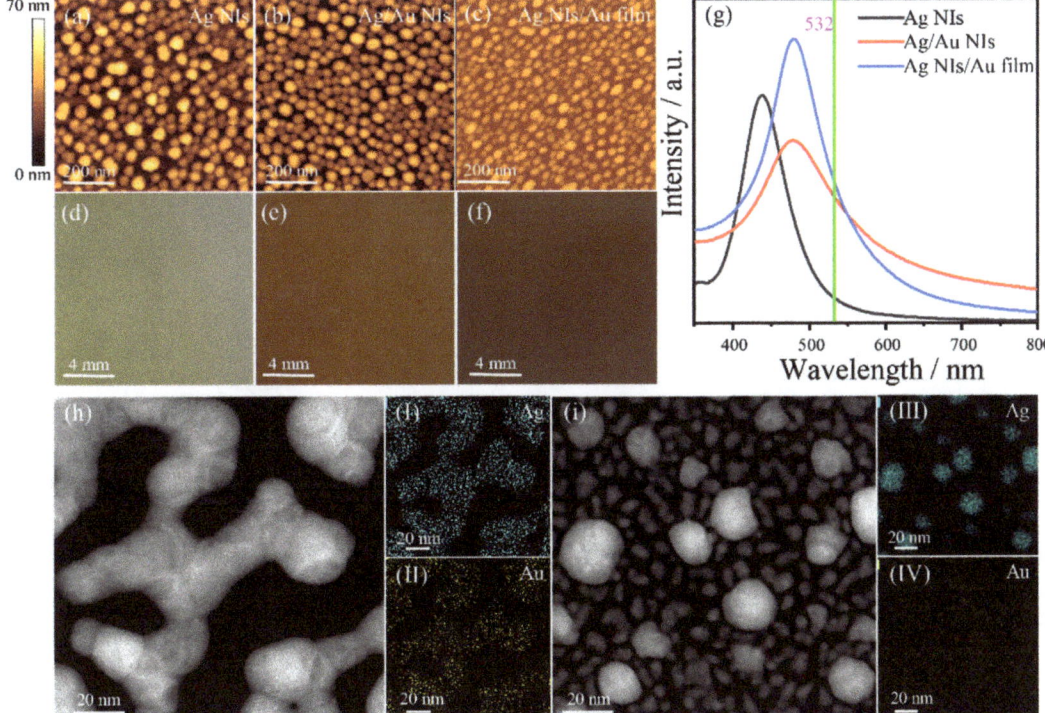

**Figure 2.** Characterization of metal NIs structures. (**a**–**c**) AFM image, (**d**–**f**) photo and (**g**) absorption spectrum of Ag NIs, Ag/Au NIs and Ag NIs/Au film. (**h**,**i**) HAADF-STEM image, and (**I**–**IV**) EDX elemental mapping of Ag/Au NIs and Ag NIs/Au film.

Plasmon thermocatalysis of NI films was investigated by observing the transformation of $NaYF_4$:$Eu^{3+}$ particles, which was obtained through co-precipitation process. As shown in Figure 3a, the SEM image shows that the product has a flower-like structure, and the overall size is about 500 nm. In the upper right corner is its tenfold magnified SEM image, more detailed sample characterization information is given in Figure S1. The plasmonic photothermal catalysis efficiency of metal NI films were studied by distributing the polycrystalline NaYF$_4$ particles uniformly on the NI films and monitoring the spectral changes in the samples under laser irradiation. Figure 3b and Figure S3 show the in situ luminescence spectra of Eu$^{3+}$-doped single sub-microparticle on the metal NI films before and after irradiation with 532 nm wavelength laser. It was found that the luminescence intensity and monochromaticity were greatly improved, and the morphology of the sub-microparticle changed from nanoflower to smoothly spherical particle, of which the image of the sample morphology was located in the upper right corner of Figure 3b. Based on the fluorescence spectra, SEM images, and previous work [20], it is suggested that the Eu$^{3+}$-doped particles after laser irradiation are single crystal spherical Y$_2$O$_3$:Eu$^{3+}$. To determine the thermal catalysis of metallic NIs, polycrystalline NaYF$_4$:Eu$^{3+}$ on a glass sheet were irradiated with laser, and the spectrum did not present any change after 15 min of irradiation with a 23 mW 532 nm laser, as shown in Figure S2. The process by which the transition occurs can be understood as follows: by laser irradiation, the LSPR of NIs is excited and the coherent plasmonic oscillations decay is formed through Landau damping, from which the hot electrons and local heat was generated. Then, these hot electrons can rapidly redistribute energy among low-energy electrons through an electron-electron scattering process. Subsequently, electrons transfer energy to the lattice through electron–phonon coupling, and the equilibrium characterized by high lattice temperature occurs within picoseconds. Due to the high thermal conductivity of NaYF$_4$ compared to the surrounding medium (air), the heat generated by the NIs dissipates through the interface with NaYF$_4$ nanoflowers through phonon–phonon interactions. Continued thermalization will eventually lead to the temperature equilibrium between NIs and NaYF$_4$ within a few nanoseconds. When sufficient heat is delivered to the lattice, NaYF$_4$ nanoflowers will begin to transform and finally recrystallize into spherical single-crystalline Y$_2$O$_3$ nanoparticles with minimal specific surface area.

**Figure 3.** (a) SEM images of NaYF$_4$:Eu$^{3+}$, the inset is the SEM image after ten times magnification; (b) In situ luminescence spectra of Eu$^{3+}$-doped sub-microparticle before and after laser irradiation (23 mW), and inserted SEM images show initial and transformed sub-microparticles, respectively.

The dynamic process of crystal conversion driven by NIs plasmon can be studied by monitoring the fluorescence emission of Eu$^{3+}$ while controlling the laser irradiation time, and all NIs were obtained by annealing in air, at 200 °C. Firstly, the plasmonic thermocatalytic rate of metal NIs can be regulated by varying the laser radiation power. Figure 4a shows the dependence of irradiation time and power required for the transformation of NaYF$_4$ particle into single crystal Y$_2$O$_3$. As the power increases from 5.0 mW to 22.5 mW, the time required for the crystal transition decreases. For the same irradiation power, Ag

NIs, Ag/Au NIs and Ag NIs/Au film catalyze crystal transformation with different rates. Compared with Ag NIs, the crystal transition rate of Ag/Au NIs and Ag NIs/Au film are 4 times and 10 times higher, respectively. At low temperature of 213 K (−60 °C), the metal NI films still present a strong LSP thermal effect, and the generated local heat can also drive the crystal transition. As shown in Figure 4b, when the temperature decreases from 20 °C to −60 °C, although the time required for crystal transformation increases, the transformation can also occur in a short time. In particular, compared with Ag NIs, the stability of the LSPR of the prepared Ag/Au composite NIs system is much better than Ag NIs. As shown in Figure 4c, the transition time also depends on the storage time of the NI films in air. As the storage time increases, the crystal transition time driven by Ag NIs increases. After the one-month storage, the transition time is 17 times that of the original. This is due to the active properties of Ag, which is easily oxidized and reduces the thermal effects of LSPR. However, for Ag/Au NIs and Ag NIs/Au films, the LSPR thermal effect is much more stable even after being stored in the air for one month. It is proved that the Ag/Au composite system can overcome the weakness of the Ag NIs through depositing the Au on the surface of the Ag, which brings the obvious improvement in the catalytic efficiency and stability.

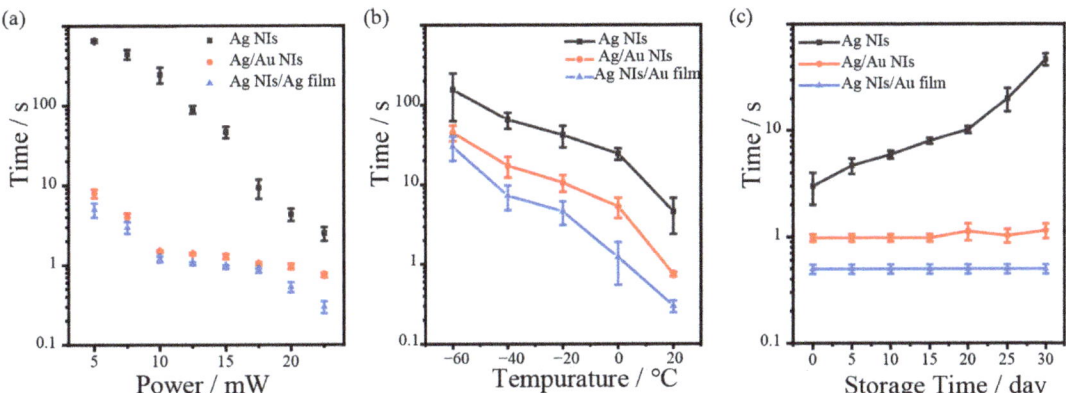

**Figure 4.** (a) Irradiation power dependence on transformation time; (b) transformation time at low temperature environment with a laser power of 22.5 mW at 532 nm; (c) NI film storage time dependence on the transformation time in air, and the irradiation power is 20 mW laser.

Since the LSPR of metal NIs depends on the geometric properties island size and gap, a series of Ag/Au composite structures were prepared at different annealing temperatures to study the LSPR thermocatalytic efficiency. As shown in Figure 5, the annealing temperature in the range of RT to 400 °C was selected to study the effect of plasmon driven crystal transformation of Ag/Au NIs (a) and Ag NIs/Au films (b). With the increase in temperature, the island particle size and gap increases. Under 532 nm and 22.5 mW laser radiation, the transformation rate of LSPR photothermal drive crystal first increases and then decreases for NIs of I to V in Figure 5, and the fastest transformation was obtained with the NIs III. At the same temperature, the transformation time of Ag NIs/Au film is faster than that of the Ag/Au NIs, as shown in Figure 5c,d. These results suggest that the photothermal catalytic efficiency of LSPR can be controlled by changing the size and gap of plasmonic NIs.

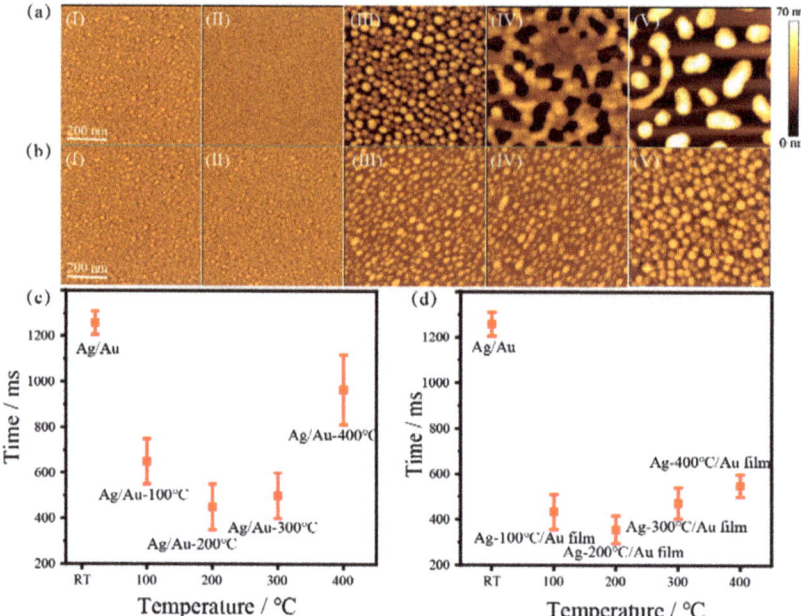

**Figure 5.** (**a**) Ag/Au NIs and (**b**) Ag NIs/Au films' AFM images of smooth (**I**) and annealing at 100 °C, 200 °C, 300 °C, 400 °C (**II–V**), respectively; (**c**,**d**) are the corresponding nano-island-driven crystal transition times, respectively.

## 4. Conclusions

The plasmonic thermocatalytic effect of metal NIs was investigated by monitoring the transformation rate of the polycrystalline $NaYF_4:Eu^{3+}$ particle to single crystalline $Y_2O_3:Eu^{3+}$ particle. Compared with Ag NIs, Ag/Au NIs and Ag NIs/Au film composite systems present better LSPR stability and thermocatalytic efficiency. It is found that the LSPR thermocatalytic efficiency of metal NIs can be controlled by changing the laser radiation power and morphology of the NIs. Even at a low temperature, NIs can still generate enough amount of heat to drive the crystal transformation. The current study can provide a simple and fast way for the application of Ag plasmon catalysis, which may enable researchers to break the limitation of traditional methods to obtain crystal transition.

**Supplementary Materials:** The following supporting information can be downloaded at: https://www.mdpi.com/article/10.3390/nano12172927/s1, Figure S1: (a) Transmission electron microscopy (TEM) characterization of $NaYF_4:Eu^{3+}$ nanoflowers; (b) XRD pattern of $NaYF_4:Eu^{3+}$ and standard pattern of cubic $NaYF_4$. Figure S2: In situ luminescence spectra of $NaYF_4:Eu^{3+}$ on glass irradiated with 532nm laser irradiation. Figure S3: Evolution of the luminescence spectra for crystal transformation Ag NIs/Au thin film substrate, under 532 nm laser irradiation (22.5 mW at the sample).

**Author Contributions:** Conceptualization, X.W., C.Z. and Z.Z.; Methodology, C.Z.; formal analysis, Z.F. and L.Y.; investigation, X.W. and X.Z.; writing—original draft preparation, X.W., writing—review and editing, X.Z., J.L. and H.Z.; supervision, J.L.; project administration, Z.Z.; funding acquisition, Z.Z. and H.Z. All authors have read and agreed to the published version of the manuscript.

**Funding:** This work was funded by the National Key R&D Program of China (Grant No. 2020YFA0211300; 2021YFA1201500), the National Natural Science Foundation of China (Nos. 92050112, 92150110, 12074237 and 12004233, 12104366), the Fundamental Research Funds for Central Universities (GK202201012 and GK202103018, 2022JQ-041), and The Young Talent fund of University Association for Science and Technology in Shaanxi and Xi'an (Grant 20220518, Grant 095920221311).

**Institutional Review Board Statement:** Not applicable.

**Informed Consent Statement:** Not applicable.

**Data Availability Statement:** Data are contained within the article or Supplementary Materials.

**Conflicts of Interest:** The authors declare no conflict of interest.

## References

1. Halas, N.J.; Lal, S.; Chang, W.S.; Link, S.; Nordlander, P. Plasmons in strongly coupled metallic nanostructures. *Chem. Rev.* **2011**, *111*, 3913–3961. [CrossRef] [PubMed]
2. Aslam, U.; Rao, V.G.; Chavez, S.; Linic, S. Catalytic conversion of solar to chemical energy on plasmonic metal nanostructures. *Nat. Catal.* **2018**, *1*, 656–665. [CrossRef]
3. Meng, X.G.; Liu, L.Q.; Ouyang, S.X.; Xu, H.; Wang, D.F.; Zhao, N.Q.; Ye, J.H. Nanometals for solar-to-chemical energy conversion: From semiconductor-based photocatalysis to plasmon-mediated photocatalysis and photo-thermocatalysis. *Adv. Mater.* **2016**, *28*, 6781–6803. [CrossRef] [PubMed]
4. Linic, S.; Aslam, U.; Boerigter, C.; Morabito, M. Photochemical transformations on plasmonic metal nanoparticles. *Nat. Mater.* **2015**, *14*, 567–576. [CrossRef] [PubMed]
5. Linic, S.; Christopher, P.; Ingram, D.B. Plasmonic-metal nanostructures for efficient conversion of solar to chemical energy. *Nat. Mater.* **2011**, *10*, 911–921. [CrossRef]
6. Kazuma, E.; Kim, Y. Mechanistic studies of plasmon chemistry on metal catalysts. *Angew. Chem. Int. Ed. Engl.* **2019**, *58*, 4800–4808. [CrossRef]
7. Kale, M.J.; Avanesian, T.; Christopher, P. Direct photocatalysis by plasmonic nanostructures. *ACS Catal.* **2014**, *4*, 116–128. [CrossRef]
8. Rycenga, M.; Cobley, C.M.; Zeng, J.; Li, W.; Moran, C.H.; Zhang, Q.; Qin, D.; Xia, Y. Controlling the synthesis and assembly of silver nanostructures for plasmonic applications. *Chem. Rev.* **2011**, *111*, 3669–3712. [CrossRef] [PubMed]
9. Xie, W.; Schlucker, S. Hot electron-induced reduction of small molecules on photorecycling metal surfaces. *Nat. Commun.* **2015**, *6*, 7570. [CrossRef] [PubMed]
10. Wang, J.L.; Ando, R.A.; Camargo, P.H.C. Investigating the plasmon-mediated catalytic activity of AgAu nanoparticles as a function of composition: Are two metals better than one? *ACS Catal.* **2014**, *4*, 3815–3819. [CrossRef]
11. Da Silva, A.G.; Rodrigues, T.S.; Correia, V.G.; Alves, T.V.; Alves, R.S.; Ando, R.A.; Ornellas, F.R.; Wang, J.; Andrade, L.H.; Camargo, P.H. Plasmonic nanorattles as next-generation catalysts for surface plasmon resonance-mediated oxidations promoted by activated oxygen. *Angew. Chem. Int. Ed. Engl.* **2016**, *55*, 7111–71115. [CrossRef] [PubMed]
12. Agrawal, A.; Cho, S.H.; Zandi, O.; Ghosh, S.; Johns, R.W.; Milliron, D.J. Localized surface plasmon resonance in semiconductor nanocrystals. *Chem. Rev.* **2018**, *118*, 3121–3207. [CrossRef] [PubMed]
13. Mascaretti, L.; Dutta, A.; Kment, S.; Shalaev, V.M.; Boltasseva, A.; Zboril, R.; Naldoni, A. Plasmon-enhanced photoelectrochemical water splitting for efficient renewable energy storage. *Adv. Mater.* **2019**, *31*, 1805513. [CrossRef] [PubMed]
14. Wang, C.; Astruc, D. Nanogold plasmonic photocatalysis for organic synthesis and clean energy conversion. *Chem. Soc. Rev.* **2014**, *43*, 7188–7216. [CrossRef] [PubMed]
15. Wu, K.; Rindzevicius, T.; Stenbæk, M.; Mogensen, K.; Xiao, S.; Boisen, A. Plasmon resonances of Ag capped Si nanopillars fabricated using mask-less lithography. *Opt. Express.* **2015**, *23*, 12965–12978. [CrossRef] [PubMed]
16. Chung, T.; Lee, Y.; Ahn, M.S.; Lee, W.; Bae, S.I.; Hwang, C.; Jeong, K.H. Nanoislands as plasmonic materials. *Nanoscale* **2019**, *11*, 8651–8664. [CrossRef] [PubMed]
17. Loza, K.; Heggen, M.; Epple, M. Synthesis, Structure, Properties, and Applications of Bimetallic Nanoparticles of Noble Metals. *Adv. Funct. Mater.* **2020**, *30*, 1909260. [CrossRef]
18. Kong, T.; Zhang, C.Y.; Gan, X.T.; Xiao, F.J.; Li, J.P.; Fu, Z.K.; Zhang, Z.L.; Zheng, H.R. Fast transformation of a rare-earth doped luminescent sub-microcrystal via plasmonic nanoislands. *J. Mater. Chem. C* **2020**, *8*, 4338–4342. [CrossRef]
19. Zhang, C.Y.; Lu, J.B.; Jin, N.N.; Dong, L.; Fu, Z.K.; Zhang, Z.L.; Zheng, H.R. Plasmon-driven rapid in situ formation of luminescence single crystal nanoparticle. *Small* **2019**, *15*, 1901286. [CrossRef] [PubMed]
20. Zhang, C.Y.; Kong, T.; Fu, Z.K.; Zhang, Z.L.; Zheng, H.R. Hot electron and thermal effects in plasmonic catalysis of nanocrystal transformation. *Nanoscale* **2020**, *12*, 8768–8774. [CrossRef] [PubMed]

Article

# Plasmon Driven Nanocrystal Transformation by Aluminum Nano-Islands with an Alumina Layer

Xilin Zhou [1], Huan Chen [1], Baobao Zhang [1], Chengyun Zhang [2,*], Min Zhang [1], Lei Xi [1], Jinyu Li [1], Zhengkun Fu [1,*] and Hairong Zheng [1]

[1] School of Physics and Information Technology, Shaanxi Normal University, Xi'an 710119, China
[2] School of Electronic Engineering, Xi'an University of Posts & Telecommunications, Xi'an 710121, China
* Correspondence: cyzhang@xupt.edu.cn (C.Z.); zkfu@snnu.edu.cn (Z.F.)

**Abstract:** The plasmonic photothermal effects of metal nanostructures have recently become a new priority of studies in the field of nano-optics. Controllable plasmonic nanostructures with a wide range of responses are crucial for effective photothermal effects and their applications. In this work, self-assembled aluminum nano-islands (Al NIs) with a thin alumina layer are designed as a plasmonic photothermal structure to achieve nanocrystal transformation via multi-wavelength excitation. The plasmonic photothermal effects can be controlled by the thickness of the $Al_2O_3$ and the intensity and wavelength of the laser illumination. In addition, Al NIs with an alumina layer have good photothermal conversion efficiency even in low temperature environments, and the efficiency will not decline significantly after storage in air for 3 months. Such an inexpensive $Al/Al_2O_3$ structure with a multi-wavelength response provides an efficient platform for rapid nanocrystal transformation and a potential application for the wide-band absorption of solar energy.

**Keywords:** surface plasmon; plasmonic photothermal; rare earth doped nanocrystal; wide-range absorption

## 1. Introduction

In recent years, there has been increasing interest in the plasmonic photothermal effects of metal nanostructures, which are widely used in a variety of applications including biomedical therapy [1–4], photocatalysis [5–8], photothermal imaging [9,10] and optofluidic technology [11,12]. Oscillations of charge carriers in plasmonic metal nanoparticles activated by resonant absorption of light are accompanied by local temperature increase due to nonradiative plasmon damping, and the heat generated is subsequently transferred to the surrounding medium; the whole process typically occurs at timescale of nanoseconds [13–15]. According to the surface plasmons decaying process, the plasmonic photothermal effects have the characteristics of efficient photothermal energy conversion, fast heat production rate, and the generation of extremely high temperatures on the nanoscale. It is important to effectively use the thermal energy from surface plasmons light absorption. In addition, plasmonic photothermal effects are strongly dependent on the intrinsic frequency of the metal nanostructures and the excitation wavelength. Metal nanostructures with narrow plasmon linewidths can only respond to a specific wavelength, forming a photothermal mode and reducing the efficiency of light utilization, which limits their applications such as solar energy harnessing [16,17].

The basic strategy to change the spectral absorption and improve the efficiency of photothermal conversion is to select suitable metal materials and modify the construction of nanostructures [18,19]. Aluminum nanomaterials possess a much wider optical range of localized surface plasmon resonances (LSPR) than gold, silver and copper, from the ultraviolet to the near-infrared region. In addition, its low cost, high natural abundance, and ease of processing make Al a sustainable plasmonic material [20,21]. Since aluminum is susceptible to oxidation and the resulting degradation of optical properties, a thin oxide layer formed

---

**Citation:** Zhou, X.; Chen, H.; Zhang, B.; Zhang, C.; Zhang, M.; Xi, L.; Li, J.; Fu, Z.; Zheng, H. Plasmon Driven Nanocrystal Transformation by Aluminum Nano-Islands with an Alumina Layer. *Nanomaterials* **2023**, *13*, 907. https://doi.org/10.3390/nano13050907

Academic Editor: Lucien Saviot

Received: 23 January 2023
Revised: 23 February 2023
Accepted: 24 February 2023
Published: 28 February 2023

**Copyright:** © 2023 by the authors. Licensee MDPI, Basel, Switzerland. This article is an open access article distributed under the terms and conditions of the Creative Commons Attribution (CC BY) license (https://creativecommons.org/licenses/by/4.0/).

rapidly on the surface of high-purity aluminum is important to protect it from further oxidation and contamination, and to improve durability. Recently, it has been reported that photothermal efficiency can be improved by fabricating aluminum micro/nano-structures to achieve anti-icing/deicing and solar water desalination [22,23]. However, the preparation of these broadband absorbing structures demands a complex nanofabrication process and often relies on sophisticated equipment, further limiting their large-scale production. It is valuable for practical applications to fabricate aluminum nanostructures with high plasmonic property by simple, rapid and low-cost technology processes.

In this work, simply prepared self-assembled Al NIs with an alumina layer are selected as a plasmonic photothermal substrate to achieve a fast transformation of rare-earth-doped nanocrystals. It has been found that Al NIs with an alumina layer can respond to several different wavelength lasers simultaneously, and they have a wide range of absorption, from visible to near-infrared light. With the increase in the thickness of the alumina heat-trapping layer, the heat utilization is further enhanced, thus significantly improving the transformation efficiency of the nanocrystals. Additionally, the $Al/Al_2O_3$ structure still produces efficient nanocrystal transformation even in low temperature environments. Additionally, the alumina layer on the Al NIs surface prevents further oxidation of the aluminum and keeps the crystal transformation efficiency stable in air for at least three months.

## 2. Materials and Methods

The substrates used are glass, which are cleaned before loading into the high vacuum coating. It consists of soaking glass substrates with piranha solution for 24 h and then ultrasonically cleaning with alcohol, acetone and deionized water for half an hour. Piranha solution is a mixture of 98% concentrated sulfuric acid and 30% hydrogen peroxide solution according to the volume ratio of 3:1. Due to its own strong oxidizing property, the solution can be used to remove organic residues from the glass substrate. $NaYF_4$:$Eu^{3+}$ was prepared by a wet-chemical method. The raw materials were NaF (98%), $Y(NO_3)_3$ (99.9%), and $Eu(NO_3)_3$ (99.9%). All reagents were purchased from Sigma–Aldrich Chemicals Co. (Shanghai, China) First, suitable stoichiometric proportions of NaF, $Y(NO_3)_3$ and $Eu(NO_3)_3$ were dissolved in a conical flask with an appropriate amount of deionized water to form a mixture, and then the complex solution was heated at 75 °C for 2 h. After the temperature was cooled down, the solution was centrifuged and washed with deionized water and ethanol twice, forming the white precipitated product. The white precipitated product was then dried at 55 °C for 12 h to obtain the target material.

To characterize the sample morphologies, scanning electron microscopy (SEM) images were obtained with a FEI-Nova NanoSEM 450 at 10 kV, and atomic force microscopy (AFM) images were obtained with a Bruker-JPK Nano Wizard Ultras (Karlsruhe, Germany). The X-ray diffraction (XRD) pattern was obtained by using a Bruker D8 Advance diffractometer (Karlsruhe, Germany). UV–Vis spectra were acquired with a PerkinElmer Lambda 950 spectrometer (Waltham, MA, USA). Conventional bright-field TEM images were obtained with a Thermo Fisher Talos F200i (Waltham, MA, USA) operated at 200 kV. In situ laser irradiation and luminescence spectra measurements were conducted with a Lab Ram HR Evolution Raman system with a 100× (NA = 0.9) objective. To avoid any crystal transformation during luminescence spectral acquisition, a low laser power (0.2 mW) was used to obtain luminescence emission.

## 3. Results and Discussion

Figure 1a illustrates the configuration of a wide-range laser wavelengths response photothermal system, which is excited by a multi-wavelength laser to produce heat for crystal transformation. Using $Al/Al_2O_3$ with a broadband plasmon resonance activity as a plasmonic heat source, the excitation of a rare-earth-doped nanocrystal placed on top with a multi-wavelength laser can lead to rapid nanocrystal transformation. Figure 1b shows the preparation process of the plasmonic photothermal system. Firstly, aluminum is deposited onto preprocessed glass substrates by thermal evaporation under a high vacuum, and then annealed at 300 °C under argon atmosphere to form aluminum nanoparticle arrays (Al NIs films). Then, Al$_2$O$_3$ layers with different thicknesses are deposited onto the Al NIs by atomic layer deposition (ALD). Finally, rare-earth-doped luminescent nanocrystals (NaYF$_4$:Eu$^{3+}$) are deposited on the Al/Al$_2$O$_3$ structure by dropwise addition.

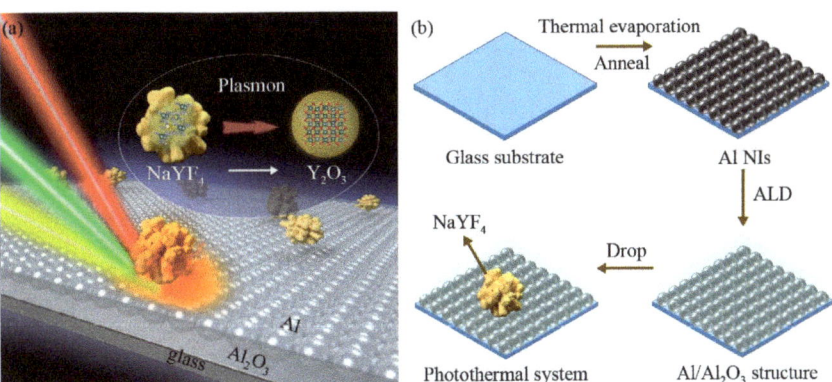

**Figure 1.** (**a**) Schematic illustration of the fast transformation of nanocrystals driven by the plasmonic photothermal system with a multi-wavelength response. (**b**) Preparation process of the plasmonic photothermal system.

Compared with lithography and chemical synthesis, metal evaporation followed by short-term thermal annealing is a simple, fast and low-cost method and allows the formation of well-separated nanostructures over large areas [24]. Figure 2a is the SEM image of aluminum deposited on Si substrate with annealing at 300 °C, where Al NPs are well separated to form an array of aluminum nanoparticles. As shown in the AFM image in Figure 2b, the root mean square value of the surface roughness of Al NIs is 3.741 nm. Due to the AFM tip-sample convolution effect, the nanostructured particles in the AFM image are larger in size compared to the SEM. During AFM scanning, the tip-sample convolution effect is caused by the geometrical interactions between the tip and surface features being imaged. The tip-sample convolution effect is one of the main causes of AFM artifacts, owing to the finite sharpness and characteristic geometry of tips [25]. In addition, aluminum films of the same deposited thickness are annealed at different temperatures, and as seen from the optical and AFM images the average size and gap of the particles do not change significantly with increasing annealing temperatures (Figure S1). It is the spontaneous oxidation of aluminum that produces the alumina layer, which improves thermal stability, effectively increasing the heat resistance of the aluminum film and limiting the migration rate of aluminum atoms during the annealing treatment [26,27]. In the subsequent experiments, Al NIs films annealed at 300 °C are selected as the substrates. Next, the Al$_2$O$_3$ layer is uniformly deposited on the surface of the Al NIs. The SEM image of the cross-section of the Al/Al$_2$O$_3$ structure shows that the thickness of Al$_2$O$_3$ is approximately 52 nm (Figure 2c). More detailed information of the Al/Al$_2$O$_3$ structure is shown in Figure S2.

Rare-earth-doped luminescent nanocrystals (NaYF$_4$:Eu$^{3+}$) are prepared by a wet-chemical method (see Materials and Methods). The SEM image in Figure 3a shows that the NaYF$_4$:Eu$^{3+}$ particles are in the shape of flowers, with uniform size and good dispersion. A single nanoflower particle size of about 500 nm is shown in the TEM image in Figure 3b, and the suitable particle size and good dispersion are favorable for the subsequent photothermal study of the surface plasmon induced nanocrystal transformation of the metal nano-islands films. The high-resolution TEM image in Figure 3c shows that the NaYF$_4$:Eu$^{3+}$ nanoflower consists of many small grains, and the crystallinity of the sample is poor because the sample was not treated at high temperature during the synthesis process. In addition, the XRD pattern shows that NaYF$_4$ is a cubic phase, and the intensity and sharpness of the diffraction spectrum of the sample are slightly lower, indicating the poor crystallinity of the sample (Figure S3). The EDX elemental mapping of the nanoflower in Figure 3d shows the presence of Na, Y, F and Eu.

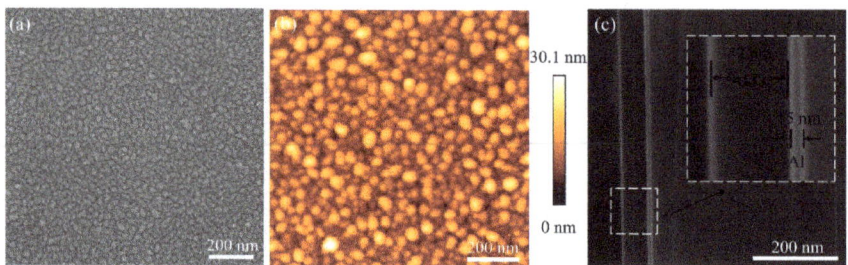

**Figure 2.** (**a**) SEM image of the Al NIs with annealing at 300 °C. (**b**) AFM image of the Al NIs annealed at 300 °C on a glass substrate. (**c**) SEM image of a cross-section of the Al/Al$_2$O$_3$ structure.

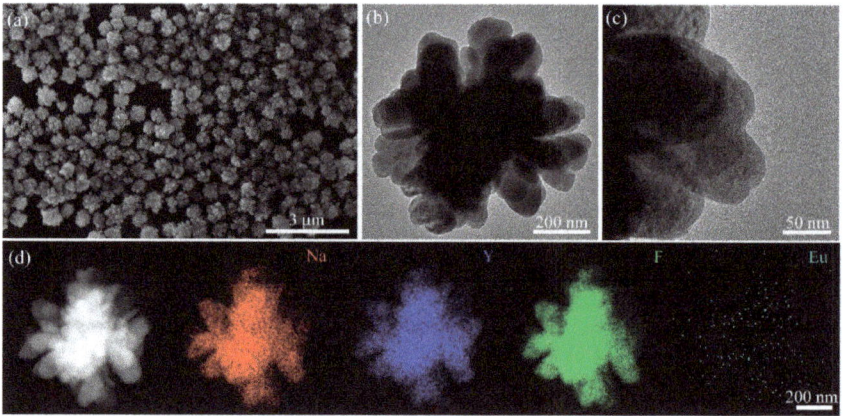

**Figure 3.** (**a**) SEM image, (**b**) TEM image, (**c**) enlarged high-resolution TEM image, (**d**) HAADF-STEM image and EDX elemental mapping of the NaYF$_4$:Eu$^{3+}$ flower-shaped nanocrystals.

The luminescence spectrum of doped Eu$^{3+}$ is used to monitor the crystal transformation of the matrix materials [28,29]. A low power (0.2 mW) 532 nm laser is used as the excitation source to obtain the luminescence spectrum, and a high power (23 mW) 976 nm laser is used as the irradiation light to drive the crystal transformation. The in situ luminescence spectra of a single nanoflower before and after 976 nm laser irradiation are shown in Figure 4a. Wide bands of luminescence centered at 590 nm, 615 nm and 700 nm with weak intensity are observed, which indicates the poor crystallinity of the initial NaYF$_4$. After laser irradiation, there is a sharp band at 610 nm with strong luminescence intensity, indicating that the nanocrystal has better crystallinity. From the inset in Figure 4a, it can be seen that the single flower-like NaYF$_4$ nanocrystal is transformed in situ into a spherical particle. The distribution of Y, O and Al elements in the product shows that the NaYF$_4$:Eu$^{3+}$ particle transformed to Y$_2$O$_3$:Eu$^{3+}$ after laser irradiation (Figure 4b). More details on the product can be found in previous work [8]. Because of the short relaxation time and the high temperature of the plasmonic photothermal effects, it is difficult to detect and measure the heat production by conventional methods. On the other hand, thermal effects play an important role in the plasmon driven crystal transformation, and hot electrons play an auxiliary and synergistic role. Therefore, the irradiation time of the plasmon-driven rare-earth-doped luminescent nanocrystal transformation is used to estimate the plasmonic photothermal effects. The rate of the above crystal transformation can be easily controlled by the power of laser irradiation. As shown in Figure 4c, the irradiation time required for the transformation to $Y_2O_3$ depends on the laser power. As the laser power increases from 10 mW to 22 mW, the irradiation time decreases from approximately 1.5 s to 40 ms for 976 nm with the $Al/Al_2O_3$ structure, indicating that the plasmonic photothermal effects improve. With the increase in laser intensity, the electromagnetic field of surface plasmons can be enhanced in the same proportion, and the density and thermal effect of hot electrons caused by electromagnetic field attenuation can also increase, thus improving the transformation efficiency of the luminous crystals.

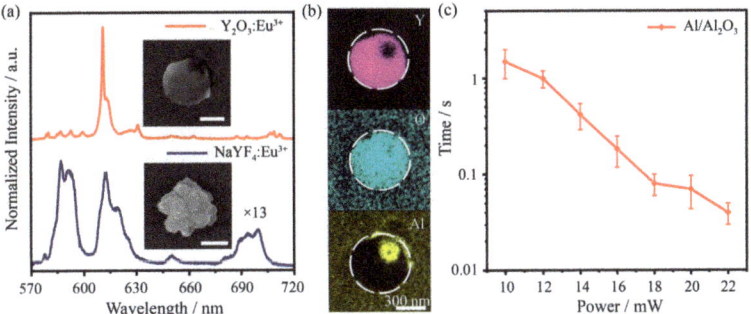

**Figure 4.** (**a**) In situ luminescence spectra of a $Eu^{3+}$-doped single nanoflower before and after 976 nm laser irradiation, and inserted SEM images show the initial and transformed nanocrystal. Scale bar, 300 nm. (**b**) EDX elemental mapping of the spherical product. (**c**) Laser power dependent irradiation time.

As shown in Figure 5a, the response of the $Al/Al_2O_3$ structure to different laser wavelengths of 532 nm, 633 nm and 976 nm is investigated, and all these laser irradiations can induce structural transformations in the crystals. In Figure 5b, by using Al NIs without an $Al_2O_3$ layer as the photothermal substrate, it is found that 532 nm, 633 nm and 976 nm laser irradiation could induce crystal structural transformation, but 532 nm and 633 nm as the irradiation laser could not induce the complete crystal transform to $Y_2O_3$. Figure 5c shows that both Al NIs and the $Al/Al_2O_3$ structure have absorption in the range of 350–1100 nm, and the intensity of the $Al/Al_2O_3$ structure is even weaker. These three different wavelengths of laser irradiation can all induce collective resonant behavior of the electrons and generate heat to drive the transformation of the rare earth nanocrystals. The introduction of alumina can effectively improve the plasmon-induced photothermal conversion efficiency.

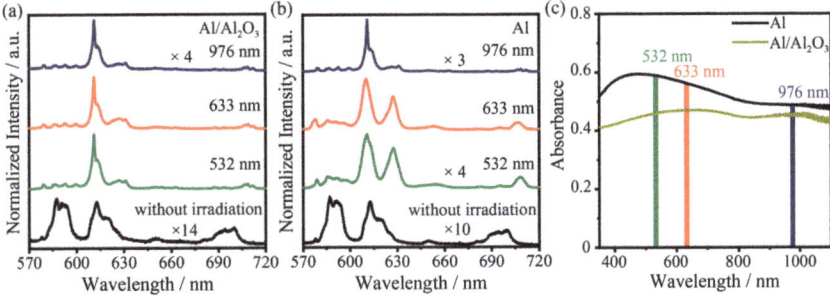

**Figure 5.** Luminescence spectra of the (**a**) $Al/Al_2O_3$ structure and (**b**) Al NIs plasmon driven nanoflower transformation after laser irradiation at 532, 633, and 976 nm. (**c**) UV–Vis spectra of Al NIs and $Al/Al_2O_3$ structure.

The difference occurs because the $Al_2O_3$ layer alters the polarizability and absorption cross-section of the aluminum nanoparticles, increasing the heat production efficiency [30,31]. On the other hand, the alumina layer has a relatively higher thermal conductivity than the glass substrate. Therefore, the heat transfer is more favorable through the alumina layer rather than through the air and the glass substrate. Figure 6a shows a heat transfer schematic diagram with and without the $Al_2O_3$ heat trapping layer. The arrow points in the direction of the heat flow. A large amount of heat generated by the aluminum nanoparticles around the sample can be transferred to it via the $Al_2O_3$ layer, reducing heat diffusion into the air and the glass substrate, leading to an increase in the temperature of the nanocrystals and an enhanced crystal transformation efficiency. In order to further investigate the dependence of the $Al_2O_3$ layer enhanced crystal transformation efficiency, the crystal irradiation time as a function of $Al_2O_3$ thickness is shown in Figure 6b. The irradiation laser used in the experiment is 976 nm (23 mW). Firstly, the required irradiation time is about 600 ms at a thickness of 5 nm $Al_2O_3$ and then decreases gradually within the thickness range from 5 to 40 nm and finally plateaus at 40 ms from 40 to 70 nm. Initially, as the thickness of the alumina increases, it enhances the heat transfer to the crystal and raises the crystal temperature rapidly; later, as the thickness further increases, the mild change in crystal temperature is not sufficient to affect the crystal transformation time. In addition, $NaYF_4:Eu^{3+}$ was placed on a smooth glass or on only a 50 nm $Al_2O_3$ layer, and the particles were irradiated with a 976 nm laser for 30 min (Figure S4). The nanoflower spectra hardly changed before and after irradiation, indicating that the heat driving the nanocrystal transformation mainly comes from the metal under laser irradiation.

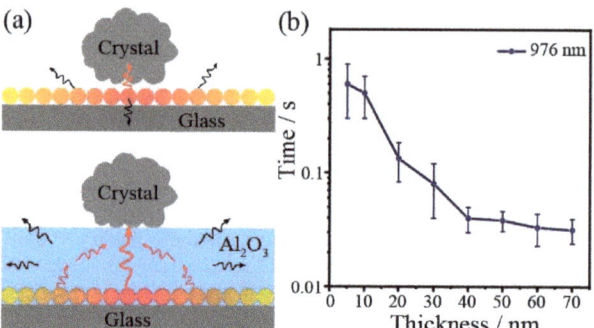

**Figure 6.** (**a**) Schematic diagram of heat transfer in the system with and without the $Al_2O_3$ heat trapping layer. (**b**) Crystal irradiation time for Al NIs as a function of $Al_2O_3$ thickness, from 5 to 70 nm.

With the $Al_2O_3$ heat-trapping layer, plasmon driven nanocrystal transformation is also realized in low temperature environments. As shown in Figure 7a, the required laser irradiation time increases from 2 s to 36 s with a decreasing temperature from 20 to −60 °C. Although a longer irradiation time is required in low temperature environments, nanocrystal transformation still occurs in an acceptably short time. In addition, both Al NIs and the $Al/Al_2O_3$ structure can be stored stably in air for at least three months. As shown in Figure 7b, the required transformation times for Al NIs with and without $Al_2O_3$ are both well stabilized. Al oxides rapidly in air and forms a thin self-limiting oxide layer. Because of the compact texture of alumina, it protects the metal inside from further oxidation, making the plasmonic photothermal effects stable. Therefore, the plasmonic photothermal system has very good heat production in low temperature environments and long-term storage stability in air, which make it more suitable for practical applications.

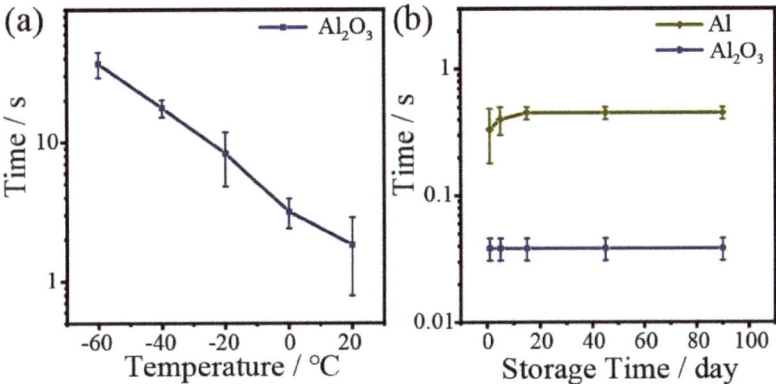

**Figure 7.** (a) Irradiation time in low-temperature environments with a laser power of 21.5 mW at 976 nm. (b) The stability of irradiation time for Al NIs and the Al/Al$_2$O$_3$ (50 nm) structure.

## 4. Conclusions

In summary, a fast crystal transformation from polycrystalline NaYF$_4$ nanoflowers to globular crystal Y$_2$O$_3$ is realized, with the self-assembled Al NIs with a thin Al$_2$O$_3$ layer as a plasmonic photothermal substrate. It has been demonstrated that the Al/Al$_2$O$_3$ structure can be excited by lasers with a wide range of wavelengths. The Al$_2$O$_3$ heat-trapping layer is shown to facilitate the coupling of heat generated near the Al nano-islands into the NaYF$_4$:Eu$^{3+}$ nanoflowers. The increased thickness of the Al$_2$O$_3$ layer enhances the efficiency of the heat transfer, resulting in a faster crystal transformation rate. In addition, the Al/Al$_2$O$_3$ structures show excellent photothermal effects even in low-temperature environments and can be preserved in air for at least three months. The low cost, simple preparation process, broad plasmon resonance activity and excellent stability make the Al/Al$_2$O$_3$ structure potentially useful in solar thermal conversion.

**Supplementary Materials:** The following supporting information can be downloaded at: https://www.mdpi.com/article/10.3390/nano13050907/s1, Figure S1: (a) Optical and (b) AFM images of the Al film without annealing (corresponding to I) and with annealing at 100 °C, 200 °C, 300 °C, 400 °C and 500 °C (corresponding to II–VI), respectively. Figure S2: (a) SEM image and (b) AFM image of the Al/Al$_2$O$_3$ structure. (c) AFM image of 50 nm Al$_2$O$_3$ later. Figure S3: XRD pattern of the as-synthesized NaYF$_4$:Eu$^{3+}$ nanoflower and the standard pattern of cubic-phase NaYF$_4$ (JCPDS No.77-2042). Figure S4: Luminescence spectra of NaYF$_4$:Eu$^{3+}$ (a) on a glass substrate without Al NIs and (b) with 50 nm Al$_2$O$_3$ deposited on a glass substrate before and after 976 nm (23 mW) laser irradiation for 30 min.

**Author Contributions:** Conceptualization, C.Z., Z.F. and H.Z.; validation, X.Z.; investigation, X.Z., H.C., B.Z., C.Z., M.Z. and L.X.; writing—original draft, X.Z.; writing—review and editing, X.Z., H.C., B.Z., C.Z., M.Z. and Z.F.; visualization, X.Z., B.Z. and J.L.; supervision, C.Z., Z.F. and H.Z. All authors have read and agreed to the published version of the manuscript.

**Funding:** This work was supported by the National Natural Science Foundation of China (Nos. U22A6005, 92150110, 92050112 and 12104366), the National Key R&D Program of China (Nos. 2020YFA0211300 and 2021YFA1201500), and the Fundamental Research Funds for Central Universities (No. GK202201012).

**Data Availability Statement:** Data are contained within the article or Supplementary Materials.

**Conflicts of Interest:** The authors declare no conflict of interest.

## References

1. Kim, M.; Lee, J.H.; Nam, J.M. Plasmonic Photothermal Nanoparticles for Biomedical Applications. *Adv. Sci.* **2019**, *6*, 1900471. [CrossRef] [PubMed]
2. Huang, P.; Lin, J.; Li, W.W.; Rong, P.F.; Wang, Z.; Wang, S.J.; Wang, X.P.; Sun, X.L.; Aronova, M.; Niu, G.; et al. Biodegradable Gold Nanovesicles with an Ultrastrong Plasmonic Coupling Effect for Photoacoustic Imaging and Photothermal Therapy. *Angew. Chem. Int. Ed.* **2013**, *52*, 13958–13964. [CrossRef]
3. Qiu, G.G.; Gai, Z.B.; Tao, Y.L.; Schmitt, J.; Kullak-Ublick, G.A.; Wang, J. Dual-Functional Plasmonic Photothermal Biosensors for Highly Accurate Severe Acute Respiratory Syndrome Coronavirus 2 Detection. *ACS Nano* **2020**, *14*, 5268–5277. [CrossRef]
4. Kumar, A.; Kim, S.; Nam, J.-M. Plasmonically Engineered Nanoprobes for Biomedical Applications. *J. Am. Chem. Soc.* **2016**, *138*, 14509–14525. [CrossRef] [PubMed]
5. Yang, Q.H.; Xu, Q.; Yu, S.H.; Jiang, H.L. Pd Nanocubes@ZIF-8: Integration of Plasmon-Driven Photothermal Conversion with a Metal-Organic Framework for Efficient and Selective Catalysis. *Angew. Chem. Int. Ed.* **2016**, *55*, 3685–3689. [CrossRef] [PubMed]
6. Meng, X.G.; Liu, L.Q.; Ouyang, S.X.; Xu, H.; Wang, D.F.; Zhao, N.Q.; Ye, J.H. Nanometals for Solar-to-Chemical Energy Conversion: From Semiconductor-Based Photocatalysis to Plasmon-Mediated Photocatalysis and Photo-Thermocatalysis. *Adv. Mater.* **2016**, *28*, 6781–6803. [CrossRef] [PubMed]
7. Kong, T.; Zhang, C.; Gan, X.; Xiao, F.; Li, J.; Fu, Z.; Zhang, Z.; Zheng, H. Fast transformation of a rare-earth doped luminescent sub-microcrystal via plasmonic nanoislands. *J. Mater. Chem. C* **2020**, *8*, 4338–4342. [CrossRef]
8. Zhang, C.; Lu, J.; Jin, N.; Dong, L.; Fu, Z.; Zhang, Z.; Zheng, H. Plasmon-Driven Rapid In Situ Formation of Luminescence Single Crystal Nanoparticle. *Small* **2019**, *15*, 1901286. [CrossRef]
9. Wu, Y.; Ali, M.R.K.; Chen, K.C.; Fang, N.; El-Sayed, M.A. Gold nanoparticles in biological optical imaging. *Nano Today* **2019**, *24*, 120–140. [CrossRef]
10. Song, J.B.; Wang, F.; Yang, X.Y.; Ning, B.; Harp, M.G.; Culp, S.H.; Hu, S.; Huang, P.; Nie, L.M.; Chen, J.Y.; et al. Gold Nanoparticle Coated Carbon Nanotube Ring with Enhanced Raman Scattering and Photothermal Conversion Property for Theranostic Applications. *J. Am. Chem. Soc.* **2016**, *138*, 7005–7015. [CrossRef]
11. Hogan, N.J.; Urban, A.S.; Ayala-Orozco, C.; Pimpinelli, A.; Nordlander, P.; Halas, N.J. Nanoparticles heat through light localization. *Nano Lett.* **2014**, *14*, 4640–4645. [CrossRef] [PubMed]
12. Liu, G.L.; Kim, J.; Lu, Y.; Lee, L.P. Optofluidic control using photothermal nanoparticles. *Nat. Mater.* **2006**, *5*, 27–32. [CrossRef] [PubMed]
13. Amendola, V.; Pilot, R.; Frasconi, M.; Marago, O.M.; Iati, M.A. Surface plasmon resonance in gold nanoparticles: A review. *J. Phys. Condens. Matter.* **2017**, *29*, 203002. [CrossRef] [PubMed]
14. Baffou, G.; Quidant, R. Thermo-plasmonics: Using metallic nanostructures as nano-sources of heat. *Laser Photonics Rev.* **2013**, *7*, 171–187. [CrossRef]
15. Chen, H.; Jiang, Z.H.; Hu, H.T.; Kang, B.W.; Zhang, B.B.; Mi, X.H.; Guo, L.; Zhang, C.Y.; Li, J.P.; Lu, J.B.; et al. Sub-50-ns ultrafast upconversion luminescence of a rare-earth-doped nanoparticle. *Nat. Photonics* **2022**, *16*, 651–657. [CrossRef]
16. Zhang, Y.; Wang, J.S.; Qiu, J.J.; Jin, X.; Umair, M.M.; Lu, R.W.; Zhang, S.F.; Tang, B.T. Ag-graphene/PEG composite phase change materials for enhancing solar-thermal energy conversion and storage capacity. *Appl. Energy* **2019**, *237*, 83–90. [CrossRef]
17. Gao, M.; Connor, P.K.N.; Ho, G.W. Plasmonic photothermic directed broadband sunlight harnessing for seawater catalysis and desalination. *Energy Environ. Sci.* **2016**, *9*, 3151–3160. [CrossRef]
18. Linic, S.; Aslam, U.; Boerigter, C.; Morabito, M. Photochemical transformations on plasmonic metal nanoparticles. *Nat. Mater.* **2015**, *14*, 567–576. [CrossRef]
19. Sousa-Castillo, A.; Comesana-Hermo, M.; Rodriguez-Gonzalez, B.; Perez-Lorenzo, M.; Wang, Z.M.; Kong, X.T.; Govorov, A.O.; Correa-Duarte, M.A. Boosting Hot Electron-Driven Photocatalysis through Anisotropic Plasmonic Nanoparticles with Hot Spots in Au-TiO$_2$ Nanoarchitectures. *J. Phys. Chem. C* **2016**, *120*, 11690–11699. [CrossRef]
20. Gerard, D.; Gray, S.K. Aluminium plasmonics. *J. Phys. D Appl. Phys.* **2015**, *48*, 184001. [CrossRef]
21. Moscatelli, A. Plasmonics. The aluminium rush. *Nat. Nanotechnol.* **2012**, *7*, 778. [CrossRef] [PubMed]
22. Li, N.; Zhang, Y.; Zhi, H.; Tang, J.; Shao, Y.; Yang, L.; Sun, T.; Liu, H.; Xue, G. Micro/nano-cactus structured aluminium with superhydrophobicity and plasmon-enhanced photothermal trap for icephobicity. *Chem. Eng. J.* **2022**, *429*, 132183. [CrossRef]
23. Zhou, L.; Tan, Y.L.; Wang, J.Y.; Xu, W.C.; Yuan, Y.; Cai, W.S.; Zhu, S.N.; Zhu, J. 3D self-assembly of aluminium nanoparticles for plasmon-enhanced solar desalination. *Nat. Photonics* **2016**, *10*, 393–398. [CrossRef]
24. Martin, J.; Plain, J. Fabrication of aluminium nanostructures for plasmonics. *J. Phys. D Appl. Phys.* **2015**, *48*, 184002. [CrossRef]
25. Shen, J.; Zhang, D.; Zhang, F.H.; Gan, Y. AFM tip-sample convolution effects for cylinder protrusions. *Appl. Surf. Sci.* **2017**, *422*, 482–491. [CrossRef]
26. Parashar, P.K.; Sharma, R.P.; Komarala, V.K. Plasmonic silicon solar cell comprised of aluminum nanoparticles: Effect of nanoparticles' self-limiting native oxide shell on optical and electrical properties. *J. Appl. Phys.* **2016**, *120*, 143104. [CrossRef]
27. Langhammer, C.; Schwind, M.; Kasemo, B.; Zoric, I. Localized surface plasmon resonances in aluminum nanodisks. *Nano Lett.* **2008**, *8*, 1461–1471. [CrossRef] [PubMed]
28. Binnemans, K. Interpretation of europium(III) spectra. *Coord. Chem. Rev.* **2015**, *295*, 1–45. [CrossRef]

29. Wang, H.; Wang, X.L.; Liang, M.S.; Chen, G.; Kong, R.M.; Xia, L.A.; Qu, F.L. A Boric Acid-Functionalized Lanthanide Metal-Organic Framework as a Fluorescence "Turn-on" Probe for Selective Monitoring of $Hg^{2+}$ and $CH_3Hg^+$. *Anal. Chem.* **2020**, *92*, 3366–3372. [CrossRef]
30. Kong, T.; Zhang, C.; Lu, J.; Kang, B.; Fu, Z.; Li, J.; Yan, L.; Zhang, Z.; Zheng, H.; Xu, H. An enhanced plasmonic photothermal effect for crystal transformation by a heat-trapping structure. *Nanoscale* **2021**, *13*, 4585–4591. [CrossRef]
31. Zhang, B.; Kong, T.; Zhang, C.; Mi, X.; Chen, H.; Guo, X.; Zhou, X.; Ji, M.; Fu, Z.; Zhang, Z.; et al. Plasmon driven nanocrystal transformation in low temperature environments. *Nanoscale* **2022**, *14*, 16314–16320. [CrossRef] [PubMed]

**Disclaimer/Publisher's Note:** The statements, opinions and data contained in all publications are solely those of the individual author(s) and contributor(s) and not of MDPI and/or the editor(s). MDPI and/or the editor(s) disclaim responsibility for any injury to people or property resulting from any ideas, methods, instructions or products referred to in the content.

Article

# Efficient Reduction Photocatalyst of 4-Nitrophenol Based on Ag-Nanoparticles-Doped Porous ZnO Heterostructure

Shali Lin, Xiaohu Mi, Lei Xi, Jinping Li, Lei Yan, Zhengkun Fu * and Hairong Zheng

School of Physics and Information Technology, Shaanxi Normal University, Xi'an 710119, China
* Correspondence: zkfu@snnu.edu.cn

**Abstract:** Oxide-supported Ag nanoparticles have been widely reported as a good approach to improve the stability and reduce the cost of photocatalysts. In this work, a Ag-nanoparticles-doped porous ZnO photocatalyst was prepared by using metal–organic frameworks as a sacrificial precursor and the catalytic activity over 4-nitrophenol was determined. The Ag-nanoparticles-doped porous ZnO heterostructure was evaluated by UV, XRD, and FETEM, and the catalytic rate constant was calculated by the change in absorbance value at 400 nm of 4-nitrophenol. The photocatalyst with a heterogeneous structure is visible, light-responsive, and beneficial to accelerating the catalytic rate. Under visible light irradiation, the heterostructure showed excellent catalytic activity over 4-nitrophenol due to the hot electrons induced by the localized surface plasmon resonance of Ag nanoparticles. Additionally, the catalytic rates of 4 nm/30 nm Ag nanoparticles and porous/nonporous ZnO were compared. We found that the as-prepared Ag-nanoparticles-doped porous ZnO heterostructure catalyst showed enhanced catalytic performance due to the synergetic effect of Ag nanoparticles and porous ZnO. This study provides a novel heterostructure photocatalyst with potential applications in solar energy and pollutant disposal.

**Keywords:** Ag nanoparticles doped porous ZnO heterostructure; photocatalysis; localized surface plasmon resonance; synergetic effect

## 1. Introduction

Semiconductor nanomaterials have been widely explored in the field of photocatalysis because of their unique electronic and optoelectronic properties. Under light irradiation, photoexcited electrons ($e^-$) and holes ($h^+$) can be generated at the conduction band (CB) and valence band (VB) if the energy absorbed by the semiconductor photocatalyst is greater than the band gap. However, due to the recombination and poor optical cross-section in the visible region, only part of the photogenerated $e^-$–$h^+$ pairs can be transferred to the surface of the photocatalyst to drive the oxidation and redox reactions, resulting in a slow reaction rate and low catalytic efficiency [1–4]. To speed up the catalytic reaction, noble metal nanoparticles have been widely used due to their strong localized surface plasmon resonance (LSPR) in the visible and near-infrared regions [5–9]. Intriguingly, the integration of a semiconductor and noble metal nanoparticles is a successful example of a plasmon-enhanced photocatalyst that can inhibit $e^-$–$h^+$ recombination, reduce activation energy, open up new reaction pathways, lower the cost, and promote the effective utilization of solar energy [10]. The hybrid nanostructures composed of a semiconductor and plasmon metal nanoparticles are mainly designed as core–shell, noble metal nanoparticles deposited on semiconductor surfaces, Janus structure, and so on [11]. According to the contact form, they can also be divided into three types: embedded, encapsulated, and isolated forms [12]. Many approaches have been studied to add metal nanoparticles to semiconductor oxides, for example, anchor metal nanoparticles on the surface of semiconductor nanostructure by a microwave polyol process, electrodeposition, photodeposition [13–15], grown semiconductor shell around the preprepared metal nanocore by reducing corresponding metal ions or cation exchange, or others [16–20]. Compared with other plasmon nanoparticles, oxide-supported Ag nanoparticles (Ag NPs) have been widely studied as a high-performance catalyst because of being easier to synthesize and having a stronger LSPR in the visible region [21,22].

Metal–organic frameworks (MOFs) are crystalline materials formed by extensive coordination of metal ions or clusters with organic linkers, characterized by high porosity, large specific surface area, and adjustable pore size. Many kinds of nanoparticles have been encapsulated in the cavities of MOFs [23,24]. Because of the instability of the organic ligand, MOFs are used as a sacrificial precursor to obtain corresponding metal oxides by thermal decomposition; fortunately, the porous structure can still be retained [25].

As one of the water pollutants, 4-nitrophenol (4-NP) has posed a serious threat to human health; therefore, it is of great importance to reduce 4-NP to 4-aminophenol (4-AP), which is less toxic [26]. One of the most popular photocatalysts for 4-NP reduction is noble metal nanoparticles doped semiconductors. The heterostructure composed of a noble metal and semiconductor can not only enhance light absorption but also increase the junction interface, which is beneficial to electron transfer and accelerates the catalytic rate. In the presence of light irradiation, the electron is transferred between metal nanoparticles and semiconductors, so that the catalytic reactions are accelerated. However, the role of the hot electrons excited by the LSPR of noble nanoparticles in a heterostructure was ignored in most studies [27–31].

In this work, a new Ag NPs doped porous ZnO (Ag/p-ZnO) heterostructure material that responds to visible light was successfully synthesized by calcining $Ag^+$-doped MOFs. The catalytic activity and mechanism of the heterostructure under dark, UV, and visible light were analyzed. Four mixing kinds of 4 nm/30 nm Ag NPs and porous and nonporous-ZnO (p and non-p-ZnO, respectively) over 4-NP were compared. The results showed that in addition to the synergistic effect of the heterostructure, the LSPR of Ag NPs excited by visible light also has a strong influence on the electron transfer efficiency from $NaBH_4$ to 4-NP. These findings help to understand the importance of plasmon catalysis from noble nanoparticles in the heterostructure and design of a photocatalyst with a reasonable structure and effective absorption of visible light.

## 2. Experimental

### 2.1. Materials

In this study, we used zinc nitrate hexahydrate ($Zn(NO_3)_2 \cdot 6H_2O$), zinc acetate ($Zn(CH_3COO)_2$), sodium hydroxide (NaOH), adenine ($C_5H_5N_5$), 4,4'-stilbenedicarboxylicacid ($C_{16}H_{12}O_4$), N, N-dimethylformamide (DMF), purchased from Sigma-Aldrich, St. Louis, MO, USA. Silver nitrate ($AgNO_3$), sodium borohydride ($NaBH_4$), 4-nitrophenol (4-NP), and hexadecyl trimethyl ammonium chloride (CTAC), purchased from Sinopharm Chemical Reagent Co., Ltd., Shanghai, China. All of the reagents and solvents were used without further purification.

### 2.2. Materials Sample Preparation

The process for preparing the Ag/p-ZnO heterostructure is shown in Figure 1a. The MOFs were prepared via a solvothermal reaction according to the previously reported method with minor modifications [32]. To obtain more uniform nanospheres, except for the reaction carried out under magnetic stirring at 85 °C for 5 h in the water bath, the other conditions remain unchanged. The Ag/p-ZnO heterostructure were prepared according to the previously reported method with minor modifications [33]. The as-prepared MOFs' aqueous solution was soaked in $AgNO_3$ solution and continuously stirred for 3 h at room temperature; in the meanwhile, $Me_2NH_2^+$ was replaced by $Ag^+$ via a simple cation exchange process as well as the strong interactions between $Ag^+$ and the nitrogen atoms of the adenine linkers [34,35]. $Ag^+$-doped MOF composites were collected by centrifugation, then the excess $Ag^+$ on the surface of MOFs was washed away using DI water and dried at 60 °C for 12 h again. Finally, the $Ag^+$ was reduced into Ag NPs, MOFs were thermally decomposed into p-ZnO at 500 °C in air for 2 h at a heating rate of 5 °C/min, and the powder of the Ag/p-ZnO composite was obtained.

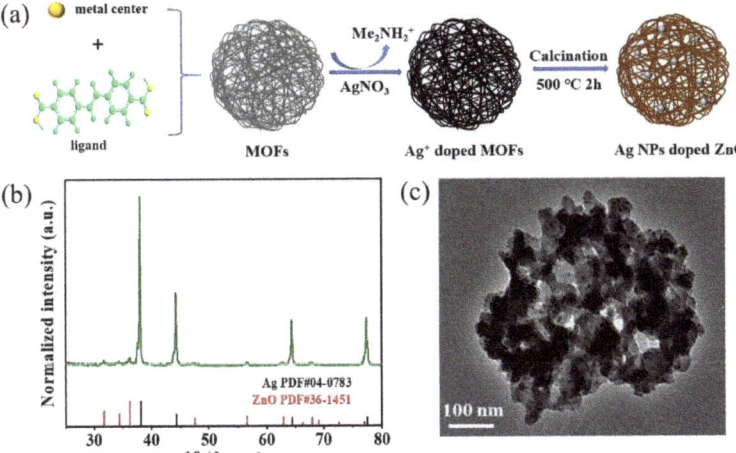

**Figure 1.** Ag-doped porous ZnO. (**a**) The synthesis procedures of Ag$^+$ doped MOFs via cation exchange and Ag/p-ZnO heterostructure by calcining. (**b**) XRD patterns of Ag/p-ZnO. (**c**) TEM image of sample Ag/p-ZnO.

First, 4 nm Ag NPs were synthesized by mixing an aqueous solution of CTAC (10 mL, 0.1 M) and AgNO$_3$ (250 mL, 0.01 M); then, the solution became dark brown when NaBH$_4$ (60 µL, 0.1 M) was added under rapid stirring. The 4 nm Ag NPs were obtained by continuously stirring for 2 min. We synthesized 30 nm Ag NPs according to the previously reported method [36]. The non-p-ZnO was prepared via a solvothermal reaction. The aqueous solution of Zn(CH$_3$COO)$_2$ (27.2 mL, 0.2 mol/L) was added to a beaker flask, then a NaOH aqueous solution (7.8 mL, 2.8 mol/L) was added drop by drop under continuous magnetic stirring for 10 min. After that, the sample was transferred into a Teflon-lined stainless-steel autoclave kept at 180 °C for 24 h, cooled to room temperature, then washed with DI water 2~3 times.

2.3. Structural and Optical Characterization

The structure and morphology of Ag/p-ZnO were characterized by X-ray diffraction (XRD, Bruker AXS, D8 advance, Karlsruhe, Germany ad field-emission transmission electron microscope (FETEM, J-2100, JEOL, Chiyoda, Japan); the UV–vis absorption spectrum was measured by a Perkin Elmer Lambda 950 spectrometer (Perkin Elmer, Lambda 950, Shanghai, China).

2.4. Catalytic Reduction of 4-NP

Typically, the catalyst was added to a mixture of 1 mL 4-NP (0.4 mM) and 1 mL NaBH$_4$ (40 mM) aqueous solution and immediately shaken homogeneously, and then transferred to a cuvette. The absorption spectra were collected from 300 to 500 nm. The corresponding mass and volume are shown in Table S1. For the blank experiment, 1 mL of 4-NP was mixed with 1 mL of NaBH$_4$ without catalyst, and the Ag NPs solution was replaced by the same volume of DI water when 5.705 mL. A conventional 3 W UV and visible LED with 3 W was used as the light source.

## 3. Results and Discussion

### 3.1. X-ray Diffraction and Field Emission Transmission Electron Microscope

The XRD patterns indicated that the Ag/p-ZnO heterostructures are highly crystalline (Figure 1b). All diffraction peaks could be attributed to hexagonal ZnO (JCPDS PDF NO.38-1451) or cubic Ag (JCPDS PDF NO.04-0783). The FETEM image of sample Ag/p-ZnO (Figure 1c) showed that the Ag NPs were embedded in p-ZnO.

### 3.2. Catalytic Performance

The catalytic performance of the prepared sample Ag/p-ZnO was studied for the reduction hydrogenation reaction of 4-NP using a NaBH$_4$-assisted reducing agent in an aqueous solution, as shown in Figure 2a. In the absence of a catalyst, electrons cannot directly transfer from NaBH$_4$ to 4-NP; therefore, they do not react and show a yellow-green color [37]. Once the catalyst was added, 4-NP reduced to 4-AP; at the same time, the yellow-green color gradually faded away and the intensity of 4-NP absorption spectra at 400 nm drastically decreased. As shown in Figure 2b, the Ag/p-ZnO sample showed the catalytic ability for 4-NP reduction under natural light, with digital photos of color changes in the insert.

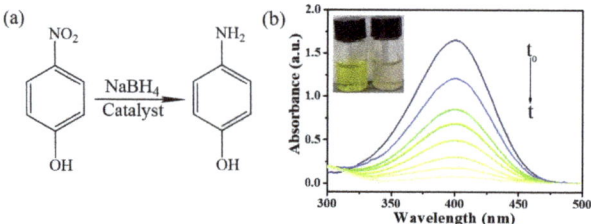

**Figure 2.** Catalytic reaction of 4-NP to 4-AP. (a) The catalytic reaction equation. (b) Time-dependent UV–Vis spectra showing gradual reduction of 4-NP over the sample Ag/p-ZnO collected in the cuvette under natural light, with digital photos of color changes before (left) and after (right) the reaction inserted.

To study the influence of light on the catalytic rate, the reaction was performed in the dark, under UV, and with visible light irradiation. The kinetic equation for the reduction can be written as $Ln(C_t/C_0) = Ln(A_t/A_0) = -k_{app}t$, where $C_t/A_t$ and $C_0/A_0$ represent the concentration/absorbance of 4-NP at time t and $t_0$, respectively; therefore, the rate constant $k_{app}$ (min$^{-1}$) can be obtained by the slope of $Ln(C_t/C_0)$ versus the reaction time (min) [38]. In Figure 3, according to the change of the slope, it can be observed that regardless of being in the dark or under light irradiation, the Ag/p-ZnO heterostructure showed the catalytic ability for 4-NP reduction and relatively high catalytic activity under visible light irradiation. Additionally, the slope exhibits little difference whether in the dark or under UV light irradiation.

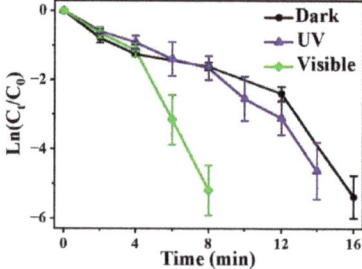

**Figure 3.** Plots of $Ln(C_t/C_0)$ versus the reaction time for the reduction of 4-NP over Ag/p-ZnO in the dark, and under UV and visible light irradiation.

## 3.3. Catalytic Mechanism

It is thought that there are two pathways for electron transfer between Ag NPs and p-ZnO under light irradiation, as shown in Figure 4. The wavelength of the irradiated light determines which component of the heterostructure is excited. Under UV light, the electrons transfer from p-ZnO to Ag NPs, while under visible light, the electrons transfer occurs from Ag NPs to p-ZnO. Under UV light, the photons are absorbed by p-ZnO, then $e^-$ and $h^+$ are generated from CB and VB, respectively. The photoexcited electrons transfer from CB to Ag NPs due to the Schottky barrier at the interface; as a result, the recombination of $e^-$–$h^+$ pairs is effectively prevented. The Ag NPs serve as sinks and promote charge separation. Under visible light, due to the strong LSPR effect from Ag NPs, a large number of photoexcited hot electrons are produced on the surface of Ag NPs, which overcome the Schottky barrier during the decay of the LSPR and transfer from Ag NPs to CB [15,39]. Based on the above experimental results, it can be seen that visible light plays a more important role in promoting the catalytic rate than that under UV light. The results showed that the hot electrons induced by the LSPR of Ag NPs more easily transfer between Ag NPs and p-ZnO, and the catalytic rate is improved.

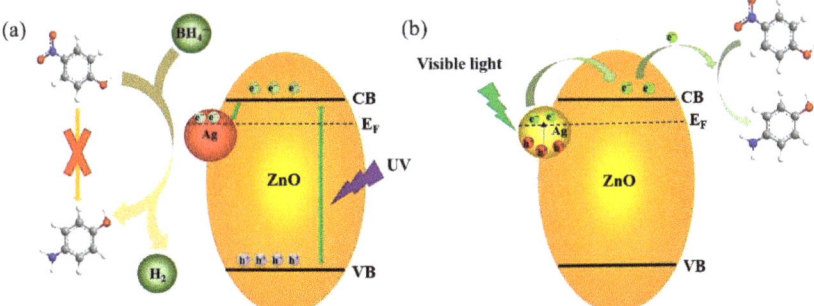

**Figure 4.** Schematic diagram of photocatalytic mechanism Ag/p-ZnO heterostructure under (**a**) UV and (**b**) visible light.

Without light, the proposed catalytic mechanism is as follows: both 4-NP and $NaBH_4$ are adsorbed on the surface of Ag/p-ZnO, and $NaBH_4$ in an aqueous solution reacts with hydroxyl-containing substances such as water, slowly releasing $H_2$, which is dissociated into polar hydrogen $H^{\delta-}$ and $H^{\delta+}$ by Ag NPs. $H^{\delta-}$ forms Ag-H on the surface of Ag NPs, which can reduce $-NO_2$ to $-NH_2$ [40].

The effect of the content of Ag NPs in a Ag/p-ZnO heterostructure on the reduction rate was examined. The content of Ag NPs in Ag/p-ZnO heterostructure could be altered by changing the concentration of $AgNO_3$, as shown by the different colors of powder in Figure S1. They were labeled as p-ZnO, Ag/p-ZnO-1 to -5. The XRD patterns are shown in Figure S2. Due to the ratio of silver to adenine, which could induce the transformation of MOFtoMOF, p-ZnO was surrounded by Ag NPs [34]. The UV–Vis absorption spectra (Figure S3) of p-ZnO showed a sharp absorption peak at about 370 nm due to excitonic absorption [41], while a broad absorption peak at 420–800 nm was due to the surface plasmon resonance (SPR) of Ag NPs with uneven size [42]. The absorption peak of p-ZnO gradually decreased with the increase in Ag NP loading. For samples Ag/p-ZnO-4 and Ag/p-ZnO-5, the Ag NPs absorption peak became a diagonal line, while the p-ZnO absorption peak gradually decreased. The most likely explanation for such changes is that Ag NPs were aggregated.

As the heterostructure showed a high catalytic rate under visible light, the following catalytic experiment was conducted under natural light. The catalytic activity of the heterostructure was optimized, as shown in Figure 5. The catalytic conversion of 4-NP over Ag/p-ZnO heterostructure under natural light was shown in Figure S5. The catalytic rate and conversion are shown in Table S2. Sample Ag/p-ZnO-1 showed excellent catalytic activity with a rate of about 0.482 min$^{-1}$ and conversion of 99%. However, we observed that the content of Ag NPs was not directly proportional to the reaction rate. For example, Ag/p-ZnO-4 and Ag/p-ZnO-5 samples had higher Ag NPs content, but the catalytic rate is reduced by about four to five times. Compared with the Ag/p-ZnO sample, the p-ZnO sample had a lower catalytic rate, so we concluded that a small amount of Ag NPs doping will greatly improve the reaction rate.

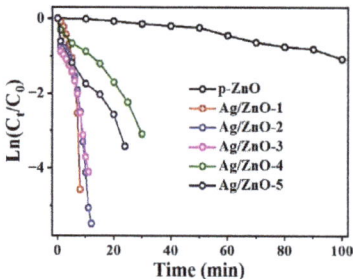

**Figure 5.** Plots of Ln($C_t/C_0$) versus the reaction time for the reduction of 4-NP over the Ag/p-ZnO heterostructure under natural light.

In addition to the wavelength of radiation light and the content of Ag NPs, the contact form between Ag NPs and ZnO also affected the catalytic rate. Comparative experiments were conducted under natural light, as shown in Figures 6 and S6. The mixture of p-ZnO and 30 nm Ag NPs showed higher catalytic activity than that of non-p-ZnO and 4 nm Ag NPs. However, despite the Ag NPs being attached to the surface of p/non-p-ZnO, their catalytic rates were not as fast as that of the Ag/p-ZnO heterostructure. p-ZnO had a large specific surface area and possessed more active sites; additionally, the embedded structure provided more contact interface between Ag NPs and p-ZnO, forming separated "islands" at the heterointerface [22]. We concluded that the excellent catalytic activity of the Ag/p-ZnO heterostructure is due to the Schottky barriers formed between Ag NPs and p-ZnO. The adequate contact surface between Ag NPs and p-ZnO not only facilitates electron transfer but also significantly enhances the catalytic performance of 4-NP.

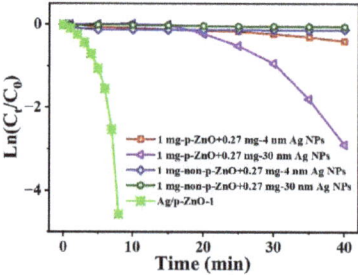

**Figure 6.** Plots of Ln($C_t/C_0$) versus the reaction time for the reduction of 4-NP over different catalysts under natural light.

## 4. Conclusions

In summary, a new visible-light responsive Ag/p-ZnO heterostructure photocatalyst was successfully synthesized using MOFs as the sacrificial precursor. The excellent catalytic performance for 4-NP indicates that the light-excited hot electrons from the LSPR can greatly accelerate the electron transfer from NaBH$_4$ to 4-NP. The rate constant of the heterostructure over 4-NP can reach about 0.482 min$^{-1}$ and is 40 times faster than that of p-ZnO under natural light. By comparing the catalytic activity of the heterostructure with that of four mixing kinds of 4 nm/30 nm Ag NPs and p/non-p-ZnO, we propose that the catalytic rate is affected by direct contact forms. Increasing the junction interface as much as possible between Ag NPs and p-ZnO is not only favorable for the catalytic reaction, but reduces the aggregation of Ag NPs. Our findings help with understanding the electron transfer process of heterostructure and designing a low-cost photocatalyst that is more environmentally friendly.

**Supplementary Materials:** The following supporting information can be downloaded at: https://www.mdpi.com/article/10.3390/nano12162863/s1. Figure S1: Digital photos of p-ZnO with different Ag NPs loading; Figure S2: XRD patterns of Ag/p-ZnO with different Ag NPs loading; Figure S3: This is a figure. Schemes follow the same formatting. The UV-visible absorption spectra of Ag/p-ZnO with different Ag NPs loading; Figure S4: Time-dependent UV-vis spectra showing gradual reduction of 4-NP over Ag/p-ZnO collected at 1 min intervals continuously under natural light; Figure S5: Catalytic conversion of 4-NP over Ag/p-ZnO heterostructure under natural light; Figure S6: Comparative mixture catalysts collected at 1 min intervals continuously under natural light. (a) 1 mg-p-ZnO + 0.27 mg 4 nm Ag NPs, (b) 1 mg-p-ZnO + 0.27 mg 30 nm Ag NPs, (c) 1 mg-non-p-ZnO + 0.27 mg 4 nm Ag NPs, (d)1 mg-non-p-ZnO + 0.27 mg 30 nm Ag NPs; Table S1: Sample mass and concentration of reaction solution required for catalytic reaction; Table S2: The rate constants kapp ($min^{-1}$) and conversion rate calculated from Figures 5 and S5.

**Author Contributions:** Conceptualization, S.L.; Data curation, X.M. and L.X.; Formal analysis, S.L., X.M. and L.X.; Methodology, S.L.; Project administration, Z.F. and H.Z.; Supervision, Z.F.; Writing—original draft, S.L.; Writing—review & editing, J.L., L.Y., Z.F. and H.Z. All authors have read and agreed to the published version of the manuscript.

**Funding:** This work was funded by the National Key R&D Program of China (grant nos. 2021YFA1201500; 2020YFA0211300), the National Natural Science Foundation of China (nos. 92150110, 92050112, 12074237, and 12004233), and the Fundamental Research Funds of Central Universities (GK202201012 and GK202103018).

**Data Availability Statement:** The data are available on reasonable request from the corresponding author.

**Conflicts of Interest:** The authors declare no conflict of interest.

## References

1. Jiang, J.; Li, H.; Zhang, L.Z. New Insight into Daylight Photocatalysis of AgBr@Ag: Synergistic Effect between Semiconductor Photocatalysis and Plasmonic Photocatalysis. *Chem.-A Eur. J.* **2012**, *18*, 6360–6369. [CrossRef] [PubMed]
2. Liu, J.; Wang, Y.; Ma, J.; Peng, Y.; Wang, A. A review on bidirectional analogies between the photocatalysis and antibacterial properties of ZnO. *J. Alloy. Compd.* **2019**, *783*, 898–918. [CrossRef]
3. Kisch, H. Semiconductor PhotocatalysisMechanistic and Synthetic Aspects. *Angew. Chem.-Int. Ed.* **2013**, *52*, 812–847. [CrossRef] [PubMed]
4. Dhiman, P.; Rana, G.; Kumar, A.; Sharma, G.; Vo, D.V.N.; Naushad, M. ZnO-based heterostructures as photocatalysts for hydrogen generation and depollution: A review. *Environ. Chem. Lett.* **2022**, *20*, 1047–1081. [CrossRef]
5. Kim, Y.; Smith, J.G.; Jain, P.K. Harvesting multiple electron-hole pairs generated through plasmonic excitation of Au nanoparticles. *Nat. Chem.* **2018**, *10*, 763–769. [CrossRef]
6. Christopher, P.; Xin, H.L.; Linic, S. Visible-light-enhanced catalytic oxidation reactions on plasmonic silver nanostructures. *Nat. Chem.* **2011**, *3*, 467–472. [CrossRef]
7. He, R.; Wang, Y.C.; Wang, X.Y.; Wang, Z.T.; Liu, G.; Zhou, W.; Wen, L.P.; Li, Q.X.; Wang, X.P.; Chen, X.Y.; et al. Facile synthesis of pentacle gold-copper alloy nanocrystals and their plasmonic and catalytic properties. *Nat. Commun.* **2014**, *5*, 1–10. [CrossRef]
8. Zhang, Z.L.; Zhang, C.Y.; Zheng, H.R.; Xu, H.X. Plasmon-Driven Catalysis on Molecules and Nanomaterials. *Acc. Chem. Res.* **2019**, *52*, 2506–2515. [CrossRef]
9. Gao, W.Q.; Liu, Q.L.; Zhao, X.L.; Cui, C.; Zhang, S.; Zhou, W.J.; Wang, X.N.; Wang, S.H.; Liu, H.; Sang, Y.H. Electromagnetic induction effect induced high-efficiency hot charge generation and transfer in Pd-tipped Au nanorods to boost plasmon-enhanced formic acid dehydrogenation. *Nano Energy* **2021**, *80*, 105543. [CrossRef]
10. Meng, X.G.; Liu, L.Q.; Ouyang, S.X.; Xu, H.; Wang, D.F.; Zhao, N.Q.; Ye, J.H. Nanometals for Solar-to-Chemical Energy Conversion: From Semiconductor-Based Photocatalysis to Plasmon-Mediated Photocatalysis and Photo-Thermocatalysis. *Adv. Mater.* **2016**, *28*, 6781–6803. [CrossRef]
11. Jiang, R.; Li, B.; Fang, C.; Wang, J. Metal/Semiconductor Hybrid Nanostructures for Plasmon-Enhanced Applications. *Adv. Mater.* **2014**, *26*, 5274–5309. [CrossRef] [PubMed]
12. Zhang, X.M.; Chen, Y.L.; Liu, R.S.; Tsai, D.P. Plasmonic photocatalysis. *Rep. Prog. Phys.* **2013**, *76*, 46401. [CrossRef] [PubMed]

13. Bian, J.-C.; Yang, F.; Li, Z.; Zeng, J.-L.; Zhang, X.-W.; Chen, Z.-D.; Tan, J.Z.Y.; Peng, R.-Q.; He, H.-Y.; Wang, J. Mechanisms in photoluminescence enhancement of ZnO nanorod arrays by the localized surface plasmons of Ag nanoparticles. *Appl. Surf. Sci.* **2012**, *258*, 8548–8551. [CrossRef]
14. Vaiano, V.; Matarangolo, M.; Murcia, J.J.; Rojas, H.; Navio, J.A.; Hidalgo, M.C. Enhanced photocatalytic removal of phenol from aqueous solutions using ZnO modified with Ag. *Appl. Catal. B-Environ.* **2018**, *225*, 197–206. [CrossRef]
15. Liu, H.R.; Hu, Y.C.; Zhang, Z.X.; Liu, X.G.; Jia, H.S.; Xu, B.S. Synthesis of spherical Ag/ZnO heterostructural composites with excellent photocatalytic activity under visible light and UV irradiation. *Appl. Surf. Sci.* **2015**, *355*, 644–652. [CrossRef]
16. Meir, N.; Plante, I.J.L.; Flomin, K.; Chockler, E.; Moshofsky, B.; Diab, M.; Volokh, M.; Mokari, T. Studying the chemical, optical and catalytic properties of noble metal (Pt, Pd, Ag, Au)-Cu2O core-shell nanostructures grown via a general approach. *J. Mater. Chem. A* **2013**, *1*, 1763–1769. [CrossRef]
17. Li, P.; Wei, Z.; Wu, T.; Peng, Q.; Li, Y.D. Au-ZnO Hybrid Nanopyramids and Their Photocatalytic Properties. *J. Am. Chem. Soc.* **2011**, *133*, 5660–5663. [CrossRef]
18. Tan, L.L.; Yu, S.J.; Jin, Y.R.; Li, J.M.; Wang, P.P. Inorganic Chiral Hybrid Nanostructures for Tailored Chiroptics and Chirality-Dependent Photocatalysis. *Angew. Chem.-Int. Ed.* **2021**, *61*, e202112400.
19. Li, W.Q.; Wang, G.Z.; Li, G.H.; Zhang, Y.X. "Ship-in-a-Bottle" Approach to Synthesize Ag@hm-SiO2 Yolk/Shell Nanospheres. *Chin. J. Chem. Phys.* **2015**, *28*, 611–616. [CrossRef]
20. Khanal, B.P.; Pandey, A.; Li, L.; Lin, Q.; Bae, W.K.; Luo, H.; Klimov, V.I.; Pietryga, J.M. Generalized Synthesis of Hybrid Metal-Semiconductor Nanostructures Tunable from the Visible to the Infrared. *ACS Nano* **2012**, *6*, 3832–3840. [CrossRef]
21. Liao, G.F.; Fang, J.S.; Li, Q.; Li, S.H.; Xu, Z.S.; Fang, B.Z. Ag-Based nanocomposites: Synthesis and applications in catalysis. *Nanoscale* **2019**, *11*, 7062–7096. [CrossRef] [PubMed]
22. Liao, G.; Gong, Y.; Zhong, L.; Fang, J.; Zhang, L.; Xu, Z.; Gao, H.; Fang, B. Unlocking the door to highly efficient Ag-based nanoparticles catalysts for NaBH4-assisted nitrophenol reduction. *Nano Res.* **2019**, *12*, 2407–2436. [CrossRef]
23. Lu, G.; Li, S.Z.; Guo, Z.; Farha, O.K.; Hauser, B.G.; Qi, X.Y.; Wang, Y.; Wang, X.; Han, S.Y.; Liu, X.G.; et al. Imparting functionality to a metal-organic framework material by controlled nanoparticle encapsulation. *Nat. Chem.* **2012**, *4*, 310–316. [CrossRef] [PubMed]
24. Li, D.D.; Yu, S.H.; Jiang, H.L. From UV to Near-Infrared Light-Responsive Metal-Organic Framework Composites: Plasmon and Upconversion Enhanced Photocatalysis. *Adv. Mater.* **2018**, *30*, 1707377. [CrossRef] [PubMed]
25. Chen, Y.Z.; Zhang, R.; Jiao, L.; Jiang, H.L. Metal-organic framework-derived porous materials for catalysis. *Coord. Chem. Rev.* **2018**, *362*, 1–23. [CrossRef]
26. Zhang, M.; Long, H.; Liu, Q.; Sun, L.B.; Qi, C.X. Synthesis of stable and highly efficient Au@ZIF-8 for selective hydrogenation of nitrophenol. *Nanotechnology* **2020**, *31*, 485707. [CrossRef]
27. Bao, Z.H.; Yuan, Y.; Leng, C.B.; Li, L.; Zhao, K.; Sun, Z.H. One-Pot Synthesis of Noble Metal/Zinc Oxide Composites with Controllable Morphology and High Catalytic Performance. *Acs Appl. Mater. Interfaces* **2017**, *9*, 16417–16425. [CrossRef]
28. Seh, Z.W.; Liu, S.H.; Zhang, S.Y.; Shah, K.W.; Han, M.Y. Synthesis and multiple reuse of eccentric Au@TiO2 nanostructures as catalysts. *Chem. Commun.* **2011**, *47*, 6689–6691. [CrossRef]
29. Abdel-Fattah, T.M.; Wixtrom, A.; Zhang, K.; Cao, W.; Baumgart, H. Highly Uniform Self-Assembled Gold Nanoparticles over High Surface Area ZnO Nanorods as Catalysts. *ECS J. Solid State Sci. Technol.* **2014**, *3*, M61–M64. [CrossRef]
30. Li, B.X.; Hao, Y.G.; Shao, X.K.; Tang, H.D.; Wang, T.; Zhu, J.B.; Yan, S.L. Synthesis of hierarchically porous metal oxides and Au/TiO2 nanohybrids for photodegradation of organic dye and catalytic reduction of 4-nitrophenol. *J. Catal.* **2015**, *329*, 368–378. [CrossRef]
31. Lei, M.; Wu, W.; Yang, S.L.; Zhang, X.G.; Xing, Z.; Ren, F.; Xiao, X.H.; Jiang, C.Z. Design of Enhanced Catalysts by Coupling of Noble Metals (Au,Ag) with Semiconductor SnO2 for Catalytic Reduction of 4-Nitrophenol. *Part. Part. Syst. Charact.* **2016**, *33*, 212–220. [CrossRef]
32. Gao, Z.H.; Xu, B.Y.; Fan, Y.Q.; Zhang, T.J.; Chen, S.W.; Yang, S.; Zhang, W.G.; Sun, X.; Wei, Y.H.; Wang, Z.F.; et al. Topological-Distortion-Driven Amorphous Spherical Metal-Organic Frameworks for High-Quality Single-Mode Microlasers. *Angew. Chem.-Int. Ed.* **2021**, *60*, 6362–6366. [CrossRef] [PubMed]
33. Cui, J.W.; Wu, D.P.; Li, Z.Y.; Zhao, G.; Wang, J.S.; Wang, L.; Niu, B.X. Mesoporous Ag/ZnO hybrid cages derived from ZIF-8 for enhanced photocatalytic and antibacterial activities. *Ceram. Int.* **2021**, *47*, 15759–15770.
34. Jonckheere, D.; Coutino-Gonzalez, E.; Baekelant, W.; Bueken, B.; Reinsch, H.; Stassen, I.; Fenwick, O.; Richard, F.; Samori, P.; Ameloot, R.; et al. Silver-induced reconstruction of an adenine-based metal-organic framework for encapsulation of luminescent adenine-stabilized silver clusters. *J. Mater. Chem. C* **2016**, *4*, 4259–4268. [CrossRef]
35. An, J.Y.; Geib, S.J.; Rosi, N.L. Cation-Triggered Drug Release from a Porous Zinc-Adeninate Metal-Organic Framework. *J. Am. Chem. Soc.* **2009**, *131*, 8376–8377. [CrossRef] [PubMed]
36. Zhang, C.Y.; Han, Q.Y.; Li, C.X.; Zhang, M.D.; Yan, L.X.; Zheng, H.R. Metal-enhanced fluorescence of single shell-isolated alloy metal nanoparticle. *Appl. Opt.* **2016**, *55*, 9131–9136. [CrossRef] [PubMed]
37. Ghosh, S.K.; Mandal, M.; Kundu, S.; Nath, S.; Pal, T. Bimetallic Pt-Ni nanoparticles can catalyze reduction of aromatic nitro compounds by sodium borohydride in aqueous solution. *Appl. Catal. A-Gen.* **2004**, *268*, 61–66. [CrossRef]
38. Feng, J.; Su, L.; Ma, Y.H.; Ren, C.L.; Guo, Q.; Chen, X.G. CuFe2O4 magnetic nanoparticles: A simple and efficient catalyst for the reduction of nitrophenol. *Chem. Eng. J.* **2013**, *221*, 16–24. [CrossRef]

39. Naya, S.-i.; Nikawa, T.; Kimura, K.; Tada, H. Rapid and Complete Removal of Nonylphenol by Gold Nanoparticle/Rutile Titanium(IV) Oxide Plasmon Photocatalyst. *ACS Catal.* **2013**, *3*, 903–907. [CrossRef]
40. Zhao, P.X.; Feng, X.W.; Huang, D.S.; Yang, G.Y.; Astruc, D. Basic concepts and recent advances in nitrophenol reduction by gold- and other transition metal nanoparticles. *Coord. Chem. Rev.* **2015**, *287*, 114–136. [CrossRef]
41. Shan, G.Y.; Zhong, M.Y.; Wang, S.; Li, Y.J.; Liu, Y.C. The synthesis and optical properties of the heterostructured ZnO/Au nanocomposites. *J. Colloid Interface Sci.* **2008**, *326*, 392–395. [CrossRef] [PubMed]
42. He, E.J.; Zheng, H.R.; Dong, J.; Gao, W.; Han, Q.Y.; Li, J.N.; Hui, L.; Lu, Y.; Tian, H.N. Facile fabrication and upconversion luminescence enhancement of $LaF_3:Yb^{3+}/Ln^{3+}@SiO_2$ (Ln = Er, Tm) nanostructures decorated with Ag nanoparticles. *Nanotechnology* **2014**, *25*, 045603. [CrossRef] [PubMed]

Article

# Impurity Controlled near Infrared Surface Plasmonic in AlN

Quanjiang Li [1], Jingang Wang [2], Shenghui Chen [1,*] and Meishan Wang [1,3,*]

[1] School of Physics and Optoelectronics Engineering, Ludong University, Yantai 264025, China; lqj@ldu.edu.cn
[2] College of Science, Liaoning Petrochemical University, Fushun 113001, China; jingang_wang@lnpu.edu.cn
[3] School of Integrated Circuits, Ludong University, Yantai 264025, China
* Correspondence: csh2010@163.com (S.C.); mswang1971@163.com (M.W.)

**Abstract:** In this work, we used multi-scale computational simulation methods combined with density functional theory (DFT) and finite element analysis (FEA) in order to study the optical properties of substitutional doped aluminium nitride (AlN). There was strong surface plasmon resonance (SPR) in the near-infrared region of AlN substituted with different alkali metal doping configurations. The strongest electric field strength reached $10^9$ V/m. There were local exciton and charge transfer exciton behaviours in some special doping configurations. These research results not only improve the application of multi-scale computational simulations in quantum surface plasmons, but also promote the application of AlN in the field of surface-enhanced linear and non-linear optical spectroscopy.

**Keywords:** impurity controlled; surface polarirons; plasmon

## 1. Introduction

Surface plasmons (SPs) are coherent electron collective oscillations (CEO) that propagate along the interface with electromagnetic waves, and are prevalent at the interfaces of different materials [1,2]. SPs are typically observed in nano-scale metal and non-metal interfaces. Due to the nature of the interface and the electromagnetic field mode, some SPs are limited to a small region (LSPRs) [3,4], and some SPs will propagate along various paths (PSPPs) [5,6]. These two different SPs are commonly referred to as localised surface plasmons and propagated surface plasmons, respectively. The former are widely used to enhance weak signals, such as surface-enhanced Raman spectroscopy (SERS) [7–9], tip-enhanced Raman spectroscopy (TERS) [10–13], and surface-enhanced fluorescence. The latter plays an irreplaceable role in remote photocatalyst reactions and integrated optical devices, such as miniature optical waveguides and modulators. The SPs' materials are widely used in these fields, in which SPs resonance (SPR) is distributed in ultraviolet or visible light regions. For some special applications, such as optical communication, biological detection in vivo, and photon medicine [14,15], SPR in the near-infrared band is more prevalent. However, due to the rapid development of non-linear optics, SPR also plays a very important role in some non-linear optical spectroscopy fields, such as surface-enhanced two-photon excited fluorescence (SE-TPEF) [16–18], second harmonic generation (SE-SHG) [19], and coherent anti-Stokes–Raman scattering (SE-CARS) [20]. These applications require that the intensity of femtosecond lasers be enhanced in the near-infrared region, making the development of near-infrared SPs with improved SPR properties imperative. To generate SPs, the real part of a material's complex dielectric function must be negative in a certain wavelength range, and the imaginary part should be as large as possible. Because almost all metals, especially Ag and Au [21], meet these conditions in this region, these two precious metals are often used in SPs to enhance Raman or fluorescent signals. Therefore, regulating the dielectric function of materials is very important to extend the application of SPs. Heterostructure and doping are often used to regulate a material's properties. However, obtaining a material's accurate dielectric function after doping and modification is challenging. Common methods include traditional

Drude [22–24] models and equivalent media theory [25–27]. However, the practical effects of these methods on calculating the dielectric functions of doped and low-dimensional systems are very modest. Because these situations involve quantum mechanical effects, more precise quantum mechanical methods are needed.

Density functional theory (DFT) is an ab initio quantum mechanical algorithm that does not rely on any empirical parameters [28]. It is a reliable condensed matter physics and first-principles theory that can accurately calculate the electronic structure and optical properties of materials [14]. Because exchange-correlation functionals in DFT can thoroughly describe the exchange-correlation effects between electrons in a multi-electron system, this method is universally applicable to materials composed of various elements, and is not limited to the crystal structures and components. The Finite Element Analysis (FEA) is the simulation of any given physical phenomenon using the numerical technique called Finite Element Method (FEM) [29]. AlN is a common, traditional semiconductor material similar to gallium nitride (GaN) [30–32]. AlN was first synthesised in 1877. By the 1980s, aluminium nitride was widely used in microelectronics [33–35]. The stability of doped nano-cages is evaluated through binding energy calculations [36–39]. Unlike beryllium oxide, aluminium nitride is non-toxic. Aluminium nitride is treated with metal and can replace alumina and beryllium oxide in many electronic devices. AlN's energy gap is as high as 6.2 eV, which is measured in vacuum ultraviolet reflectance.

In this work, we conducted theoretical research on the electronic structure, optical properties, and SPR characteristics of a disk array using DFT on alkali doped AlN crystals. After the Li atom replaces the Al atom, the local electric field intensity between the disks at 1300 nm is as high as $10^9$ V/m. This excellent property can be applied in many fields that require near-infrared SPR, such as biomedicine, Raman spectroscopy, non-linear optics, and non-linear surface plasmons. The multi-scale calculation method used in this work, that is, the method of analysing the surface plasmon properties of a doped system through quantum mechanical calculations, can guide theoretical research into the optical properties of materials under the same conditions.

## 2. Materials and Methods

We established an AlN hexagonal lattice using DFT calculations. Two thiophenes are used as the smallest repeating units. The periodic boundary direction is consistent with the polymer's length direction. A vacuum layer of 15 Å is present in the other two directions. The atomic centre basis set and the GGA-PBE functional [40,41] are calculated using the QuantumATK-2018.06-SP1 software package [42,43]. Full optimisation of the atomic geometry is performed until all of the components of the residual forces are less than 0.05 eV/Å and the total energy converges within $10^{-6}$ eV. The k-mesh is $7 \times 7 \times 1$ and the cut-off energy is 1200 eV. Using the same cut-off energy in the optical property calculations, LCAO (linear combination of atomic orbitals) is used for the basis set, the k-mesh increases to $15 \times 15 \times 1$, and the self-consistent field convergence limit increases to $10^{-8}$ eV. With the current calculation experience, the calculation method chosen in this paper is more accurate [29,44–46].

## 3. Results and Discussion

### 3.1. Crystal Lattice Structures

In this work, the geometries and optical properties of different alkali metal atom substitution doping AlN crystal cells are calculated via the ab initio method. The crystal structures optimised by DFT are shown in Figure 1. Alkali metals are doped in two configurations. In the first, alkali metal atoms occupy the position of N atoms in AlN crystals. As shown in Figure 1c–e, one N atom at the same position is replaced with Li, Na, and K atoms, respectively. Since the AlN crystal point group belongs to D3h, it is equivalent to substituting any N atom. After the N atom is replaced by three alkali metals, compared with the intrinsic AlN structure (Figure 1a), the structure after doping undergoes relatively large changes. The alkali metal shifts from the original N atom position and squeezes the

Al atom position. Due to the different radii of alkali metal atoms, the coordination mode of the atoms changes considerably. However, when the alkali metal atom replaces the Al atom, the crystal structure inevitably changes, as shown in Figure 1f–h. When Li atoms and Na atoms are substituted with Al atoms, the crystal structure changes only slightly, as shown in Figure 1f,g. This is because the atomic radii of the Li atoms and Na atoms are not significantly different from those of the Al atom. However, the radius of the K atom is relatively large (Figure 1h), so when the K atom replaces the Al atom, the crystal distortion is large.

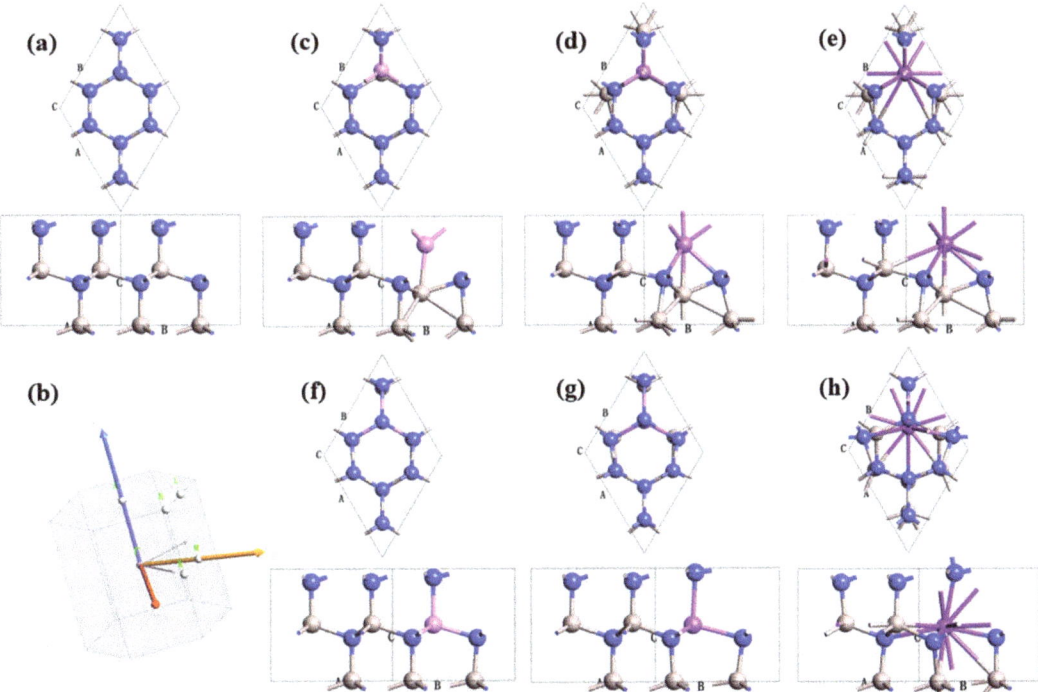

**Figure 1.** The crystal structure (**a**) and k-space schematic diagram (**b**) of intrinsic AlN. The (**c–h**) are the alkali metal substitution (N@Li, N@Na, N@K, Al@Li, Al@Na, Al@K) doping AlN crystal lattices, respectively.

*3.2. Electronic Structure*

Changes in the crystal structure can cause differences in the electronic structure. As shown in Figure 2a, the electronic band structure of AlN is a classic insulator material. The top of the valence band and the bottom value of the conduction band occur at the gamma point (Γ). The inter-band transition in AlN is a direct transition. The p orbital of the N atomic contribution plays a key role in the valence band, and the conduction band is contributed by the s and p orbitals of the Al atom. The real part of different cartesian component dielectric functions of AlN is greater than zero. Therefore, there are no strong interactions between electromagnetic waves and AlN. There is surface plasmon polarisation on the material surface where the real part of the dielectric function is less than zero and the imaginary part is as large as possible. The imaginary part of the dielectric function is greater than zero, and the real part is greater than zero. The peaks of the imaginary part of the dielectric function of AlN occur in the deep ultraviolet region. These properties limit the application of AlN on SPR, especially the optical and optoelectronic properties.

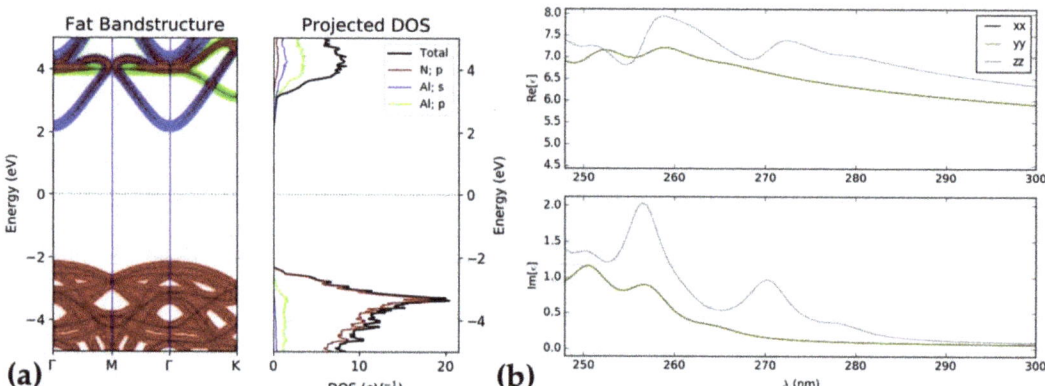

**Figure 2.** The projected electronic energy band structure, density of states (DOS) spectra (**a**) and frequency dependent isotropic dielectric function (**b**) of intrinsic AlN.

After doping, the electronic structure of AlN changes dramatically. First, replacing the Al atom position with Li or Na atoms causes the Fermi level to enter the original valence band. Near the Fermi level, no impurity level appears in the forbidden band, as shown in Figure 3a,b. However, the entry of the Fermi level into the valence band significantly affects the electronic transition. According to the DOS spectrum, there is a large state density above and below the Fermi level, so strong intra-band transitions occur. After the K atom replaces the Al atom (Figure 3c), there are no impurity levels in the forbidden band, but some p and d orbital components of the K atom are mixed into the conduction band, and the width of the forbidden band ($E_g$) decreases. Although $E_g$ decreases, it is still large for electronic transitions (~4 eV). Second, after the Li atom replaces the N atom position, the Fermi level enters the conduction band, as shown in Figure 3d. In addition, there is no small state density near the Fermi level. Therefore, in this configuration, there are no small in-band transitions. Unlike the substitution of Li atoms for Al atoms, this is an in-band transition of the conduction band, and it causes the transition in the p orbit of the Al atom. Third, after the Na and K atoms replace the N atoms, $E_g$ significantly decreases. However, the Fermi level remains in the forbidden band, and the density of states near the Fermi level is not small, as shown Figure 3e,f. Thus, there are no small inter-band transitions in these two configurations. Unlike the first five configurations, after the K atom replaces the N atom, the impurity level of the K atom appears in the forbidden band (Figure 3f), and the Fermi energy level is the K atom contribution and the Al atom contribution, respectively. Therefore, not only the inter-band transition but also the charge transfer transition occurs in this doped configuration. The metal–insulator transition depends on the relationship between the Fermi energy level and the energy band. When the conduction band or valence band of the intrinsic material overlaps or interleaves with the Fermi energy level, the metal–insulator transition will occur, and of course topological phenomena may also occur. This is also shown in the dielectric constant below. In summary, when Li atoms and Na atoms are substituted with Al atoms, intra-band transitions on N atoms in the valence band occur. When the Li atoms are substituted by N atoms, there are intra-band transitions on the Al atoms inside the conduction band. When the Na atoms and K atoms are substituted by N atoms, low-band transitions and charge transfers occur in the crystal.

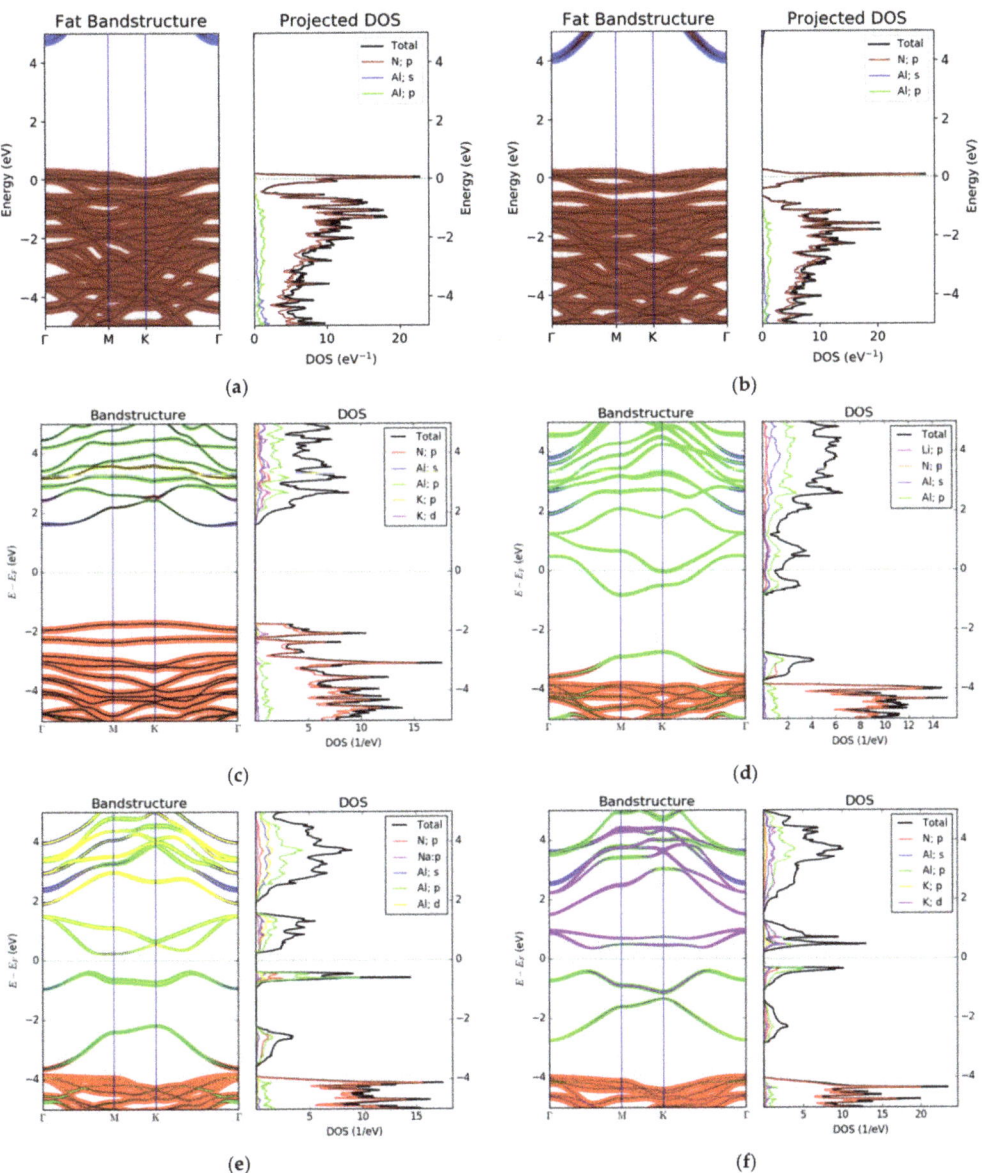

**Figure 3.** (a–f) are the electronic structures (projected energy band structure and DOS spectra) of alkali metal substitution (Al@Li, Al@Na, Al@K, N@Li, N@Na, N@K,) doping AlN, respectively.

## 3.3. Optical Properties of Doping AlN

Changes in the electronic structure can drastically affect the optical properties. According to the prior description, intra-band transitions and low-energy inter-band transitions occur in many configurations after doping. Therefore, the dielectric functions of the six configurations are calculated separately. Of course, the DFT theory does not consider temperature effects, but the ab initio molecular dynamics theory that considers temperature has little effect on the permittivity. Since AlN is a uniaxial crystal, the dielectric functions in the x and y orientations are equal. The dielectric functions of the three types of AlN and doped AlN are shown in Figure 4, respectively. Figure 4a,c and Figure 4b,d demonstrate cases in which Al atoms and N atoms are replaced, respectively. A necessary condition for the existence of a surface plasmon at a certain wavelength position is that the real part of the dielectric function is less than zero, and the imaginary part is not equal to zero. Because this is the dielectric property of metals, surface plasmons are originally found on rough precious metal surfaces. First, when Li atoms and Na atoms are substituted with Al atoms, the real part of the dielectric function in the xx orientation is less than zero in the near-infrared region, as shown in the top half of Figure 4a. However, when K atoms are substituted with Al atoms, the dielectric function has no properties in the visible and near-infrared regions that can help excite the surface plasmons. This is discussed in Section 3.2. Although the forbidden band width decreases, the inter-band transition remains in the ultraviolet region. For the z orientation shown in Figure 4c, when Li atoms replace Al atoms, the real part of the dielectric function also has a certain negative value, and the imaginary part in the corresponding region is not small. Thus, there may also be surface plasmons in the z direction. The same phenomenon also occurs after three kinds of alkali metal atom substitution types are doped at the N atom position, as shown in Figure 4b. After Li atoms are doped with N atoms, the dielectric function is less than zero in a long near-infrared region, and the value is large, reaching below −10 a.u. However, in N atom and K atom substitution doped N atoms, although there is a certain negative dielectric function at approximately 800 nm, the value is very small. This is because when the two doping configurations are used, the material has the properties of a semiconductor with a relatively small $E_g$. Therefore, in this configuration, excitons should appear in the near-infrared region, not surface plasmons.

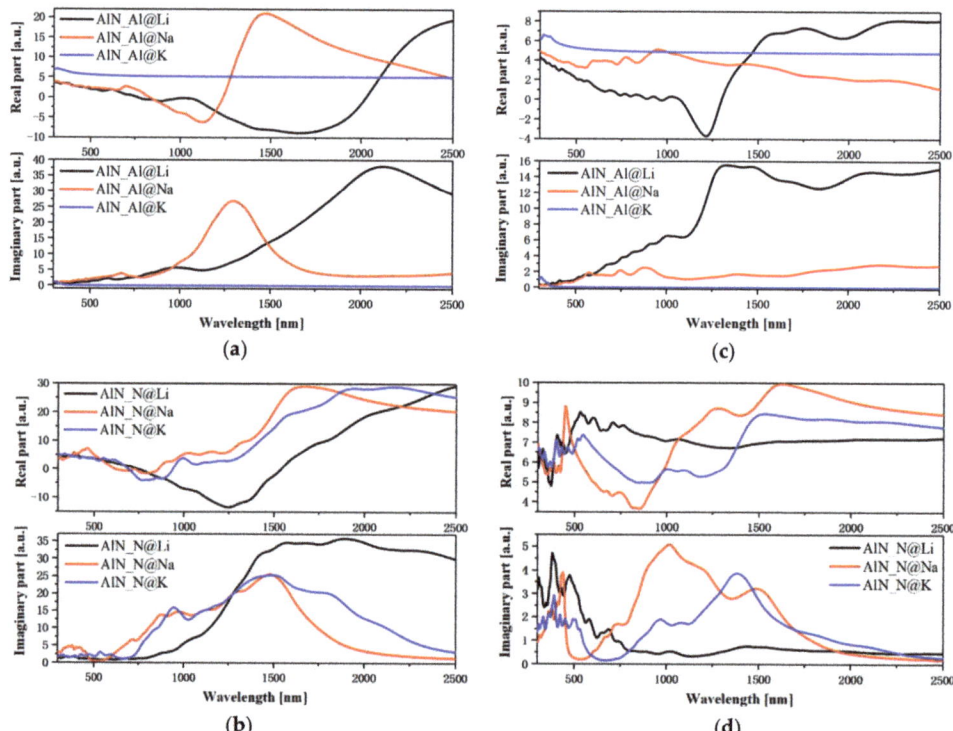

**Figure 4.** The xx and zz orientation dielectric function of alkali metal substitution (Al@Li, Al@Na, Al@K) (**a,c**) and (N@Li, N@Na, N@K,) (**b,d**) doping AlN.

The absorption spectrum can characterise the interaction strength between materials and electromagnetic waves. The absorption coefficient is defined as follows:

$$\alpha = \sqrt{\frac{\sqrt{\epsilon_1^2 + \epsilon_2^2} - \epsilon_1}{2}}, \quad (1)$$

where $\epsilon_1$ and $\epsilon_2$ are the real and imaginary parts of the dielectric function, respectively. Therefore, the absorption spectrum can be used to comprehensively analyse the effect of the complex dielectric function. When Li and Na atoms replace Al atoms, they cause strong absorption in the near-infrared region, indicating the presence of surface plasmons, as shown in Figure 5a. When K atoms are substituted, there is almost no absorption coefficient in the visible and near-infrared regions, as demonstrated in Figure 5c. However, when Li atom substitutional doping is performed, since the material also exhibits metallic properties, it also has a strong absorption peak, as shown in Figure 5b,d. After the Na atoms and K atoms replace the N atoms, the system exhibits semiconductor properties, so the absorption spectra are very close.

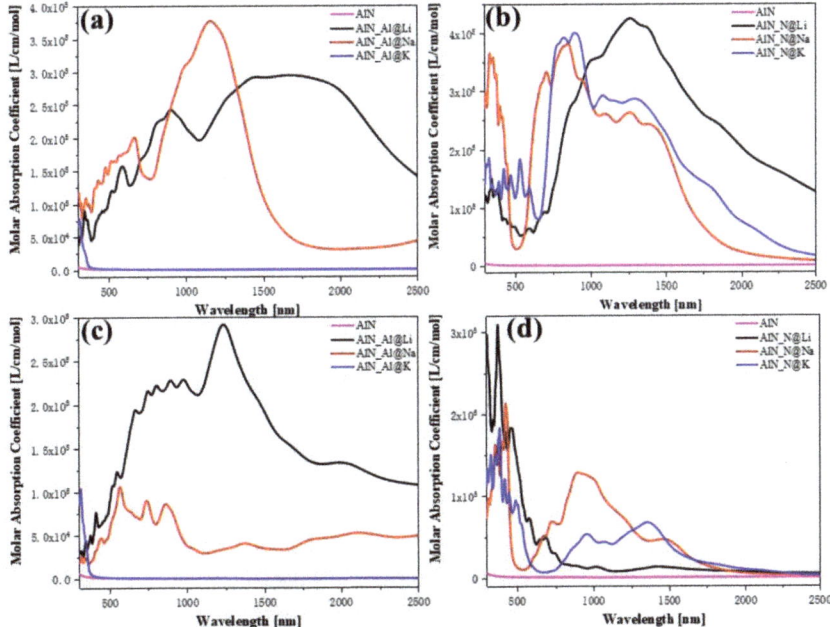

**Figure 5.** The x and z orientation absorption coefficients of alkali metal substitution (Al@Li, Al@Na, Al@K) (**a,c**) and (N@Li, N@Na, N@K,) (**b,d**) doping AlN.

### 3.4. Plasmonic and Exciton Properties

Based on the prior discussion, a surface plasmon analysis is performed for configurations that may have surface plasmons or exciton behaviour. The AlN disk array model is applied to the analysis of surface plasmons. This is because in materials with surface plasmons, there are hot spots between the disks. Therefore, in the FEA model, a disk array with a radius of 40 nm and a thickness of 20 nm with a gap of 10 nm is placed on the Si surface, as shown in Figure 6a. This is the most common array configuration. Because this work emphasises the effect of the material's properties on the surface plasmons, the fixed model size only studies the SP properties of different doping configurations. Figure 6b shows the transmission spectra of a disk array with different doping configurations. The transmission spectra peaks indicate the presence of SPR at this position. This is also where the electric field strength is strongest. The figure demonstrates that the positions of the SPR peaks are different, but they all occur in the near-infrared region. Therefore, in addition to the configuration in which the K atom replaces the Al atom, the near-infrared SPR exists in other configurations. The absolute value of the transmission spectrum of the first three configurations in Figure 6b is relatively large because these three configurations have metallic properties. The latter two configurations exhibit semiconductor properties, and the absolute value of the transmission spectrum is relatively small. Correspondingly, the electric field intensity of the first three configurations is strong and the wavelength is longer in Figure 6c, while the electric field intensity of the latter two configurations is weaker and the wavelength is shorter.

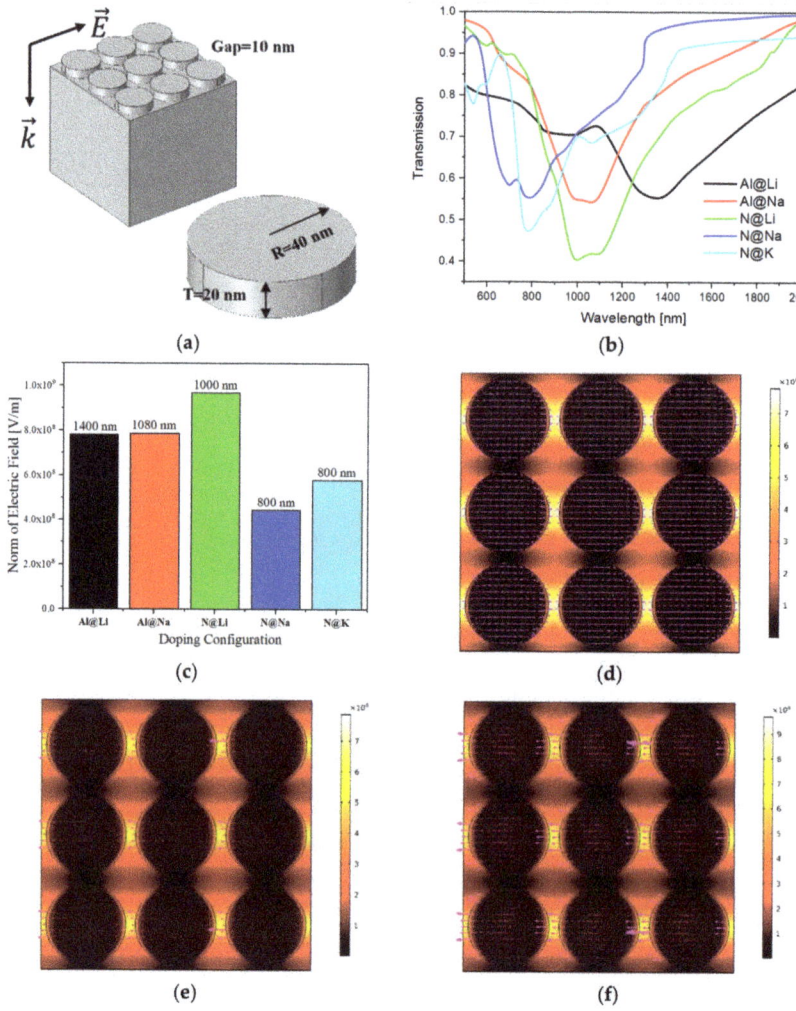

**Figure 6.** *Cont.*

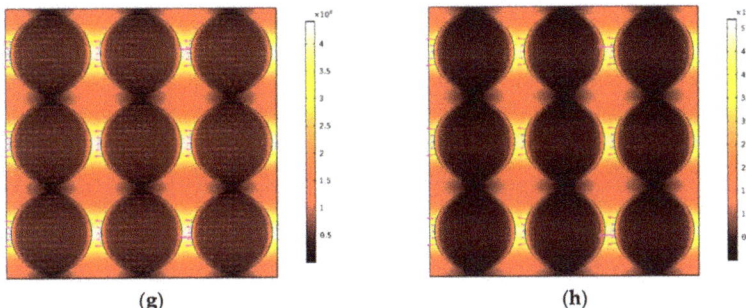

**Figure 6.** (**a**) The AlN disks array model of FEA analysis. The transmission spectra (**b**) and maximum electric field strength (**c**) of various configurations. (**d**–**h**) The top views of electric field mode of Al@Li, Al@Na, N@Li, N@Na, and N@K doping AlN, respectively. The carmine arrows are the direction of SPs polarization modes.

According to the prior analysis, there is a charge transfer exciton in the last configuration. As previously mentioned, the nature of the first three configurations and electromagnetic field interaction is SPR, so the electric field model diagram demonstrates that the locality of the electromagnetic field is very strong, as shown in Figure 6d–f. Nevertheless, the nature of the latter two configurations is exciton, so the locality is weaker (the electric field mode in the diagram in Figure 6g,h looks "brighter"). However, although the electric field modes are all similar, the polarisation directions of SPR differ significantly. When Li atoms are substituted with AL atoms, the SPR wavelength is the longest, and the polarisation direction inside the material is very stable, while the net polarisation at the hot spot is almost zero, as shown in Figure 6d. In the latter two SPR materials, the polarisation inside is partially stable, while the polarisation direction at the hot spot is opposite to the polarisation of the incident electromagnetic wave. The polarisation of the two semiconductor-type doped configuration disks is minimal. When Li atoms are doped with N atoms, the real part of the dielectric function has the smallest value, and there is a maximum electric field strength at 1000 nm, which reaches approximately $10^9$ V/m.

## 4. Conclusions

In this work, we conducted multi-scale computational simulation studies on the quantum surface plasmon and exciton properties of AlN doped with alkali metal substitutions at different positions. Depending on the doping type and location, a large SPR intensity exists in the near-infrared region. In Al@Li, Al@Na, and N@Li configurations, SPR occurs at 1400 nm and 1000 nm, respectively. At 800 nm, there are excitons in N@Na and N@K configurations, especially charge transfer excitons in N@K configurations. This paper starts from the atomic scale, simulates the optical properties of AlN doped by quantum mechanical frameworks, and analyses the surface plasmons. AlN modification improves the near-infrared surface plasmon properties (the electric field strength reaches almost $10^9$ V/m), promoting their applications in non-linear optics, CARS, and SRS surface enhancement.

**Author Contributions:** Content and design: M.W. and S.C.; Writing of manuscript: Q.L.; Software: J.W.; Drawing: Q.L. and J.W. All authors have read and agreed to the published version of the manuscript.

**Funding:** This work was supported by the National Nature Science Foundation of China (Grant No. 12104202), project ZR2019PA011 supported by the Shandong Provincial Natural Science Foundation.

**Data Availability Statement:** The data that support the findings of this study are available from the corresponding author upon reasonable request.

**Conflicts of Interest:** The authors declare no conflict of interest.

## References

1. Ritchie, R.H. Plasma losses by fast electrons in thin films. *Phys. Rev.* **1957**, *106*, 874. [CrossRef]
2. Ozbay, E. Plasmonics: Merging photonics and electronics at nanoscale dimensions. *Science* **2006**, *311*, 189–193. [CrossRef] [PubMed]
3. Hiep, H.M.; Endo, T.; Kerman, K.; Chikae, M.; Kim, D.-K.; Yamamura, S.; Takamura, Y.; Tamiya, E. A localized surface plasmon resonance based immunosensor for the detection of casein in milk. *Sci. Technol. Adv. Mater.* **2007**, *8*, 331–338. [CrossRef]
4. Downing, C.A.; Weick, G. Plasmonic modes in cylindrical nanoparticles and dimers. *Proc. R. Soc. A* **2020**, *476*, 20200530. [CrossRef]
5. Choi, S.; Park, D.; Jeong, Y.; Yun, Y.; Jeong, M.; Byeon, C.; Kang, J.; Park, Q.-H.; Kim, D. Directional control of surface plasmon polariton waves propagating through an asymmetric Bragg resonator. *Appl. Phys. Lett.* **2009**, *94*, 063115. [CrossRef]
6. Krasavin, A.; Zayats, A. Three-dimensional numerical modeling of photonic integration with dielectric-loaded SPP waveguides. *Phys. Rev. B* **2008**, *78*, 045425. [CrossRef]
7. Tong, L.; Xu, H.; Käll, M. Nanogaps for SERS applications. *MRS Bull.* **2014**, *39*, 163–168. [CrossRef]
8. Xu, H.; Bjerneld, E.J.; Käll, M.; Börjesson, L. Spectroscopy of single hemoglobin molecules by surface enhanced Raman scattering. *Phys. Rev. Lett.* **1999**, *83*, 4357. [CrossRef]
9. Xu, H.; Aizpurua, J.; Käll, M.; Apell, P. Electromagnetic contributions to single-molecule sensitivity in surface-enhanced Raman scattering. *Phys. Rev. E* **2000**, *62*, 4318. [CrossRef]
10. Zhang, Z.; Chen, L.; Sun, M.; Ruan, P.; Zheng, H.; Xu, H. Insights into the nature of plasmon-driven catalytic reactions revealed by HV-TERS. *Nanoscale* **2013**, *5*, 3249–3252. [CrossRef]
11. Zhang, Z.; Sun, M.; Ruan, P.; Zheng, H.; Xu, H. Electric field gradient quadrupole Raman modes observed in plasmon-driven catalytic reactions revealed by HV-TERS. *Nanoscale* **2013**, *5*, 4151–4155. [CrossRef] [PubMed]
12. Sun, M.; Zhang, Z.; Zheng, H.; Xu, H. In-situ plasmon-driven chemical reactions revealed by high vacuum tip-enhanced Raman spectroscopy. *Sci. Rep.* **2012**, *2*, 647. [CrossRef] [PubMed]
13. Fang, Y.; Zhang, Z.; Chen, L.; Sun, M. Near field plasmonic gradient effects on high vacuum tip-enhanced Raman spectroscopy. *Phys. Chem. Chem. Phys.* **2015**, *17*, 783–794. [CrossRef] [PubMed]
14. Li, R.; Wang, L.; Mu, X.; Chen, M.; Sun, M. Biological nascent evolution of snail bone and collagen revealed by nonlinear optical microscopy. *J. Biophotonics* **2019**, *12*, e201900119. [CrossRef]
15. Kong, L.; Wang, J.; Mu, X.; Li, R.; Li, X.; Fan, X.; Song, P.; Ma, F.; Sun, M. Porous size dependent g-C3N4 for efficient photocatalysts: Regulation synthesizes and physical mechanism. *Mater. Today Energy* **2019**, *13*, 11–21. [CrossRef]
16. Mu, X.; Cai, K.; Wei, W.; Li, Y.; Wang, Z.; Wang, J. Dependence of UV–Visible Absorption Characteristics on the Migration Distance and the Hyperconjugation Effect of a Methine Chain. *J. Phys. Chem. C* **2018**, *122*, 7831–7837. [CrossRef]
17. Mu, X.; Chen, X.; Wang, J.; Sun, M. Visualizations of Electric and Magnetic Interactions in Electronic Circular Dichroism and Raman Optical Activity. *J. Phys. Chem. A* **2019**, *123*, 8071–8081. [CrossRef]
18. Mu, X.; Wang, J.; Sun, M. Visualizations of Photoinduced Charge Transfer and Electron-Hole Coherence in Two-Photon Absorptions. *J. Phys. Chem. C* **2019**, *123*, 14123–14143. [CrossRef]
19. Valev, V. Characterization of nanostructured plasmonic surfaces with second harmonic generation. *Langmuir* **2012**, *28*, 15454–15471. [CrossRef]
20. Wu, R.-L.; Quan, J.; Tian, C.; Sun, M. Transformation from Quantum to Classical Mode: The Size Effect of Plasmon in 2D Atomic Cluster System. *Sci. Rep.* **2019**, *9*, 6641. [CrossRef]
21. Borah, R.; Verbruggen, S.W. Coupled Plasmon Modes in 2D Gold Nanoparticle Clusters and their Effect on Local Temperature Control. *J. Phys. Chem. C* **2019**. [CrossRef]
22. Shibayama, J.; Ando, R.; Nomura, A.; Yamauchi, J.; Nakano, H. Simple trapezoidal recursive convolution technique for the frequency-dependent FDTD analysis of a Drude–Lorentz model. *IEEE Photon. Technol. Lett.* **2008**, *21*, 100–102. [CrossRef]
23. Tip, A. Linear dispersive dielectrics as limits of Drude-Lorentz systems. *Phys. Rev. E* **2004**, *69*, 016610. [CrossRef] [PubMed]
24. Sabah, C.; Uckun, S. Multilayer system of Lorentz/Drude type metamaterials with dielectric slabs and its application to electromagnetic filters. *Prog. Electromagn. Res.* **2009**, *91*, 349–364. [CrossRef]
25. Choy, T.C. *Effective Medium Theory: Principles and Applications*; Oxford University Press: Oxford, UK, 2015; Volume 165.
26. Bohren, C.F. Applicability of effective-medium theories to problems of scattering and absorption by nonhomogeneous atmospheric particles. *J. Atmos. Sci* **1986**, *43*, 468–475. [CrossRef]
27. Koschny, T.; Kafesaki, M.; Economou, E.; Soukoulis, C. Effective medium theory of left-handed materials. *Phys. Rev. Lett.* **2004**, *93*, 107402. [CrossRef]
28. Szabó, B.; Babuška, I. *Finite Element Analysis*; John Wiley & Sons: Hoboken, NJ, USA, 1991.
29. Kohn, W.; Sham, L.J. Self-consistent equations including exchange and correlation effects. *Phys. Rev.* **1965**, *140*, A1133–A1138. [CrossRef]
30. Strite, S.; Morkoç, H. GaN, AlN, and InN: A review. *J. Vac. Sci. Technol. B Microelectron. Nanometer Struct. Process. Meas. Phenom.* **1992**, *10*, 1237–1266. [CrossRef]
31. Yim, W.; Paff, R. Thermal expansion of AlN, sapphire, and silicon. *J. Appl. Phys.* **1974**, *45*, 1456–1457. [CrossRef]
32. Shen, L.; Heikman, S.; Moran, B.; Coffie, R.; Zhang, N.-Q.; Buttari, D.; Smorchkova, I.; Keller, S.; DenBaars, S.; Mishra, U. AlGaN/AlN/GaN high-power microwave HEMT. *IEEE Electron Device Lett.* **2001**, *22*, 457–459. [CrossRef]

33. Ching, W.; Harmon, B. Electronic structure of AlN. *Phys. Rev. B* **1986**, *34*, 5305. [CrossRef] [PubMed]
34. Amano, H.; Sawaki, N.; Akasaki, I.; Toyoda, Y. Metalorganic vapor phase epitaxial growth of a high quality GaN film using an AlN buffer layer. *Appl. Phys. Lett.* **1986**, *48*, 353–355. [CrossRef]
35. Gerlich, D.; Dole, S.; Slack, G. Elastic properties of aluminum nitride. *J. Phys. Chem. Solids* **1986**, *47*, 437–441. [CrossRef]
36. Iqbal, J.; Ayub, K. Theoretical study of the non linear optical properties of alkali metal (Li, Na, K) doped aluminum nitride nanocages. *RSC Adv.* **2016**, *6*, 94228–94235.
37. Gueorguiev, G.K.; Goyenola, C.; Schmidt, S.; Hultman, L. CFx: A first-principles study of structural patterns arising during synthetic growth. *Chem. Phys. Lett.* **2011**, *516*, 62–67. [CrossRef]
38. Kakanakova-Georgieva, A.; Gueorguiev, G.K.; Yakimova, R.; Janzén, E. Effect of impurity incorporation on crystallization in AlN sublimation epitaxy. *J. Appl. Phys.* **2004**, *96*, 5293–5297. [CrossRef]
39. Dos Santos, R.B.; Rivelino, R.; de Brito Mota, F.; Gueorguiev, G.K.; Kakanakova-Georgieva, A. Dopant species with Al–Si and N–Si bonding in the MOCVD of AlN implementing trimethylaluminum, ammonia and silane. *J. Phys. D Appl. Phys.* **2015**, *48*, 295104. [CrossRef]
40. Perdew, J.P.; Burke, K.; Ernzerhof, M. Generalized gradient approximation made simple. *Phys. Rev. Lett.* **1996**, *77*, 3865. [CrossRef]
41. Perdew, J.P.; Chevary, J.A.; Vosko, S.H.; Jackson, K.A.; Pederson, M.R.; Singh, D.J.; Fiolhais, C. Atoms, molecules, solids, and surfaces: Applications of the generalized gradient approximation for exchange and correlation. *Phys. Rev. B* **1992**, *46*, 6671. [CrossRef]
42. Smidstrup, S.; Stradi, D.; Wellendorff, J.; Khomyakov, P.A.; Vej-Hansen, U.G.; Lee, M.-E.; Ghosh, T.; Jónsson, E.; Jónsson, H.; Stokbro, K. First-principles Green's-function method for surface calculations: A pseudopotential localized basis set approach. *Phys. Rev. B* **2017**, *96*, 195309. [CrossRef]
43. Smidstrup, S.; Markussen, T.; Vancraeyveld, P.; Wellendorff, J.; Schneider, J.; Gunst, T.; Verstichel, B.; Stradi, D.; Khomyakov, P.A.; Vej-Hansen, U.G.; et al. QuantumATK: An integrated platform of electronic and atomic-scale modelling tools. *J. Phys. Condens. Matter.* **2019**, *32*, 015901. [CrossRef] [PubMed]
44. Dev, P.; Agrawal, S.; English, N.J. Functional Assessment for Predicting Charge-Transfer Excitations of Dyes in Complexed State: A Study of Triphenylamine–Donor Dyes on Titania for Dye-Sensitized Solar Cells. *J. Phys. Chem. A* **2013**, *117*, 2114–2124. [CrossRef] [PubMed]
45. Städele, M.; Moukara, M.; Majewski, J.A.; Vogl, P.; Görling, A. Exact exchange Kohn-Sham formalism applied to semiconductors. *Phys. Rev. B* **1999**, *59*, 10031. [CrossRef]
46. Grimme, S.; Hansen, A.; Brandenburg, J.G.; Bannwarth, C. Dispersion-corrected mean-field electronic structure methods. *Chem. Rev.* **2016**, *116*, 5105–5154. [CrossRef]

Article

# Efficient Excitation and Tuning of Multi-Fano Resonances with High Q-Factor in All-Dielectric Metasurfaces

Yunyan Wang [1], Chen Zhou [1], Yiping Huo [1,*], Pengfei Cui [1], Meina Song [1], Tong Liu [1], Chen Zhao [1], Zuxiong Liao [1], Zhongyue Zhang [1] and You Xie [2]

1. Xi'an Key Laboratory of Optical Information Manipulation and Augmentation, School of Physics and Information Technology, Shaanxi Normal University, Xi'an 710062, China; yyywang@snnu.edu.cn (Y.W.); 18291927192@163.com (C.Z.); pfcui@snnu.edu.cn (P.C.); 202017027@snnu.edu.cn (M.S.); lt20211965@snnu.edu.cn (T.L.); zhaochen1999@snnu.edu.cn (C.Z.); zxliao@snnu.edu.cn (Z.L.); zyzhang@snnu.edu.cn (Z.Z.)
2. College of Science, Xi'an University of Science and Technology, Xi'an 710054, China; xieyou@xust.edu.cn
* Correspondence: yphuo@snnu.edu.cn

**Abstract:** Exciting Fano resonance can improve the quality factor (Q-factor) and enhance the light energy utilization rate of optical devices. However, due to the large inherent loss of metals and the limitation of phase matching, traditional optical devices based on surface plasmon resonance cannot obtain a larger Q-factor. In this study, a silicon square-hole nano disk (SHND) array device is proposed and studied numerically. The results show that, by breaking the symmetry of the SHND structure and transforming an ideal bound state in the continuum (BIC) with an infinite Q-factor into a quasi-BIC with a finite Q-factor, three Fano resonances can be realized. The calculation results also show that the three Fano resonances with narrow linewidth can produce significant local electric and magnetic field enhancements: the highest Q-factor value reaches 35,837, and the modulation depth of those Fano resonances can reach almost 100%. Considering these properties, the SHND structure realizes multi-Fano resonances with a high Q-factor, narrow line width, large modulation depth and high near-field enhancement, which could provide a new method for applications such as multi-wavelength communications, lasing, and nonlinear optical devices.

**Keywords:** Fano resonance; anapole resonance mode; toroidal dipole (TD) mode; quality factor (Q-factor)

## 1. Introduction

The quality factor (Q-factor) can represent the energy limiting efficiency of optical interaction between the external environment and a material. High Q-factor resonators have an important role in various fields of optical applications, such as nonlinear optics [1–3], optics sensors [4], lasers [5], and optical switches [6]. Fano resonance, produced by the destructive interference between discrete and continuous states, can form a typical sharp asymmetrical line shape and is an effective method used to achieve a high Q-factor in metasurfaces [7,8]. In the past few years, researchers have discovered that metal metasurfaces based on plasmonic resonance can generate Fano resonance [9,10]. However, a clear shortcoming is that the inherent ohmic loss of a metal metasurface is very large, which reduces the utilization rate of light energy [11]. Meanwhile, Fano resonances formed in the metal metasurface are mainly electric resonances, such as electric dipole, electric quadrupole, and high-order electrode modes, which will also cause serious radiation loss and render Q-factor enhancement difficult [12]. For both cases, because ohmic loss is an inherent property of a metal metasurface and is difficult to change, there is a need to design a special metal metasurface structure, such as the split ring structure [13,14], to excite the magnetic resonance modes and reduce energy loss, which can increase the Q-factor. Unfortunately, this method can limit the freedom of optical device design and severely limit practical applications.

Several studies have shown that a high-refractive-index all-dielectric metasurface [15,16] can significantly reduce ohmic and radiation losses, which can achieve a high Q-factor Fano resonance not matchable by a metal metasurface based on plasmon resonance. Compared to a metamaterial based on plasmonic resonance, where resonances are often dominated by electric resonance modes, all-dielectric metasurfaces can support a series of BIC and Mie resonances. Bound states in the continuum (BIC) represents an ideal physical system, which can also be referred to as capture mode. The energy is confined in the resonant cavity, showing zero radiation loss, and has the characteristic of an infinite Q-factor. By breaking the symmetry of the all-dielectric metasurfaces, a channel for energy leakage into free space is provided and BIC can be converted into quasi-BIC with a finite Q-factor, which is called the leakage mode. It is an effective method to obtain a high Q-factor in all-dielectric metasurfaces [17–20]. Mie resonance means that when electromagnetic waves are irradiated on dielectric particles, they will interact with each other to realize scattering in response to electric or magnetic fields [21,22]. This can not only generate electric resonance modes but also generate magnetic and toroidal resonance modes [23,24], thereby significantly reducing radiation loss. Currently, electric dipole (ED), magnetic dipole (MD) and toroidal dipole (TD) resonances are considered to represent the three types of fundamental modes of electromagnetics. ED resonance is generated from the separation of negative and positive charges; MD resonance is induced by the closed circulation of electric current; and TD resonance can be produced by a pair of adjacent current loops with opposite MD moments, as proposed by Zel'dovich in 1957 [25] and first experimentally observed in 2010 [26]. In an all-dielectric metasurface, the fundamental resonance is a series of magnetic resonance modes, which are useful for the confinement of the incident field within the metamaterial, leading to near-field enhancement inside the dielectric devices so that magnetic Fano resonance can be easily induced [27,28]. When the ED and TD resonances overlap, scattering the amplitude and antiphase, the non-radiating anapole mode can be generated [29,30]. Owing to the non-radiating character and efficient energy confinement of the anapole mode, a high Q-factor resonance can be achieved, which has important application in nano-lasers [31,32] and enhances nonlinear effects [33,34].

As the wide application of multiplexing methods in various devices has developed, the research focus on optical Fano resonance has also expanded from single Fano resonance to multi-Fano resonance [35,36]. Multi-Fano resonance can be used efficiently in a multi-channel biosensor and multi-band slow-light device. As recent studies have shown, multi-Fano resonance is useful in enhancing multi-band multi-harmonic generation, such as second and third harmonic generation [37], where different Fano resonances match different fundamental wavelengths and different harmonic wavelengths.

In this article, an all-dielectric silicon square-hole nano disk (SHND) array metasurface was designed and evaluated. Similar structures have been proposed previously. For example, Algori et al designed two metasurfaces based on hollow nanostructures with Q-factors of $2.5 \times 10^6$ and $1.71 \times 10^6$, but only one Fano resonance was excited [38,39], while Jeong et al. designed a dielectric metasurface structure in which two Fano resonances were excited with a Q-factor of 728 [40]. In the present investigation, by adjusting the inner square along the x-axis and breaking the symmetry of the all-dielectric metasurface, the BIC mode with an infinite Q-factor was transformed into a quasi-BIC mode with finite size, which induced triple Fano resonance. The influence of the extinction coefficient on Q-factor was investigated. Then, applying multilevel decomposition theory, it was found that the anapole, toroidal, magnetic quadrupole, and MD resonance modes were excited simultaneously. All the three Fano resonances were based on magnetic resonance and showed a high Q-factor because of weak radiative and non-radiative decay. The maximum Q-factor was able to reach 35,837 in magnitude and the modulation depths of all the Fano resonances were nearly 100%. In addition, the electromagnetic near-field enhancements could be confined inside the metasurface and were greatly enlarged by breaking the symmetry of the SHND nanostructure, which provided an effective method to modulate the localized field. Therefore, the SHND structure could operate in multi-wavelength communications, lasing, and nonlinear optical devices.

## 2. Materials and Methods

The SHND nanostructure array, shown in Figure 1, was placed on a glass substrate with a refractive index of 1.5 and completely submerged in water with a refractive index of 1.33. The thickness of the entire structure is H, the side length of the outer square with center O is L, the side length of the inner square hole with center P is W, and the distance between points O and P is defined as an asymmetry parameter g. The material that constituted the SHND structure was amorphous silicon, and the optical parameters were obtained from experimental data [41]. The initial parameters of the SHND structure were L = 600 nm, W = 200 nm, H = 100 nm, g = 0 nm, and the periodic parameters of the unit cell were Px = Py = 680 nm. The incident plane wave enters the SHND structure along the z-axis, whereas the polarization of the electric field follows the positive direction of the y-axis. All the simulation results were obtained using COMSOL Multiphysics 5.6 software, which is based on the finite element method. A three-dimensional SHND model was built in the software, and the structure was subdivided by a free triangular mesh. The degrees of freedom were 14,238 and the periodic boundary conditions were adopted as the boundary conditions. The SHND nanostructure can be prepared using a standard nano-process, as follows: firstly, deposit a thin silicon film on a silicon dioxide substrate using a low-pressure physical vapor deposition (LPCVD) method; secondly, etch rectangular holes using electron beam lithography (EBL), or a reactive ion-etching method; and finally, remove the photoresist and rinse the nanostructure with deionized water.

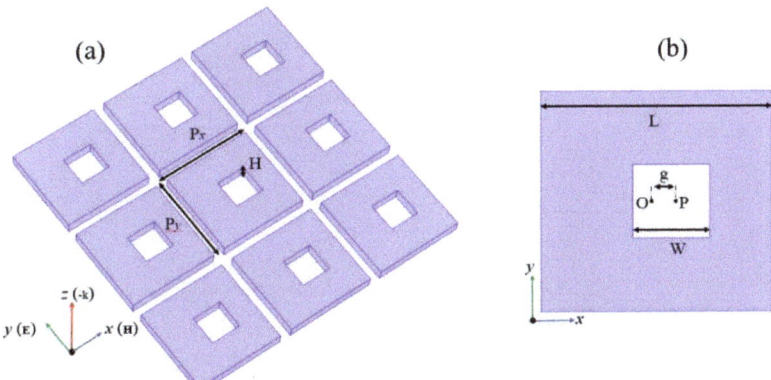

**Figure 1.** (a) Schematic of the SHND nanostructure array and the incident light polarization configuration; (b) Top view and geometric parameters of a unit cell of the SHND.

## 3. Theory

The investigation of the electromagnetic properties of the dielectric nanostructures is based on the decomposition of multipole moments in a Cartesian coordinate system. A general expression for multipoles is as follows:

Electric dipole moment:

$$\mathbf{p} = \frac{1}{j\omega} \int_V \mathbf{J}(\mathbf{r}) d\mathbf{r} \quad (1)$$

Magnetic dipole moment:

$$\mathbf{m} = \frac{1}{2v_b} \int_V [\mathbf{r} \times \mathbf{J}(\mathbf{r})] d\mathbf{r} \quad (2)$$

Toroidal dipole moment:

$$\mathbf{t} = \frac{1}{10v_b} \int_V \left[ (\mathbf{r} \cdot \mathbf{J}(\mathbf{r}))\mathbf{r} - 2r^2 \mathbf{J}(\mathbf{r}) \right] d\mathbf{r} \tag{3}$$

Electric quadrupole moment:

$$Q^e_{\alpha\beta} = \frac{1}{j2\omega} \int_V \left[ r_\alpha J_\beta + r_\beta J_\alpha - \frac{2}{3}\delta_{\alpha\beta}(\mathbf{r}\cdot\mathbf{J}(\mathbf{r})) \right] d\mathbf{r} \tag{4}$$

Magnetic quadrupole moment:

$$Q^m_{\alpha\beta} = \frac{1}{3v_b} \int_V \left\{ [\mathbf{r}\times\mathbf{J}(\mathbf{r})]_\alpha r_\beta + [\mathbf{r}\times\mathbf{J}(\mathbf{r})]_\beta r_\alpha \right\} d\mathbf{r} \tag{5}$$

where $j$ denotes the current density, $v_b$ the speed of light in the medium, $\mathbf{r}$ is the radial vector, and the subscripts for the electric and magnetic quadrupoles are $\alpha, \beta = x, y, z$. When the scatter wavelength is smaller than the incident wavelength, the higher-order terms (e.g., octupole) can be neglected.

The radiation powers for different multipole moments are given as follows:

$$I_P = \frac{2\omega^4}{3v_b^3} |\mathbf{p}|^2 \tag{6}$$

$$I_t = \frac{2\omega^6}{3v_b^5} |\mathbf{t}|^2 \tag{7}$$

$$I_m = \frac{2\omega^4}{3v_b^3} |\mathbf{m}|^2 \tag{8}$$

$$I_{Qe} = \frac{\omega^6}{5v_b^5} \left| Q^e_{\alpha\beta} \right|^2 \tag{9}$$

$$I_{Qm} = \frac{\omega^6}{40v_b^5} \left| Q^m_{\alpha\beta} \right|^2 \tag{10}$$

This multipole decomposition method allows for the identification of the contributions stemming from TD moments and hence identification of the conditions for the anapole mode excitation [42–44].

## 4. Results and Discussion

### 4.1. Excitation of Fano Resonance in the SHND Structure and Influence Factors of Q-Factor

Figure 2 shows the transmission spectra of the proposed metasurface at different asymmetry parameters. When g = 0 nm, the square hole is located at the center of the entire structure. In this case, points O and P coincide and the SHND nanostructure is completely symmetrical. There is only one asymmetric Fano resonance in the transmission spectrum. As shown in Figure 2, when g = 0 nm, there is no transmission peak except F0, which means that the Q-factor tends to infinity, showing a BIC mode. When the inner square hole moves horizontally to the right, the symmetry of the SHND nanostructures is broken; there are two obvious transmission dips near λ = 1294.9 nm and λ = 1389.5 nm, which show obvious Fano characteristics. This means that the BIC state becomes unstable with increase in the asymmetric parameter, and the breaking of symmetry provides a zero-order radiation channel for the metasurface, which leads to leakage of energy and the BIC mode is transformed into a quasi-BIC mode. The larger the asymmetry parameter, the greater the radiation loss energy and the smaller the Q-factor. When g = 40 nm, three Fano resonances appear in the transmission spectrum, which are denoted by F1, F2, and F3. Figure 2 shows that their resonance peaks are at 1182.2 nm, 1294.9 nm, and 1389.5 nm, respectively. The

Q-factor is defined by $Q = \omega_0/2\gamma$ [45,46], where $\omega_0$ is the resonance frequency and $\gamma$ is the damping loss. Figure 3 shows the variation in the Q-factor of F3 with different asymmetric parameters. When the asymmetry factor is 5 nm, the Q-factor can reach 35,837. With increase in the asymmetry parameter g, the Q-factor decreases rapidly.

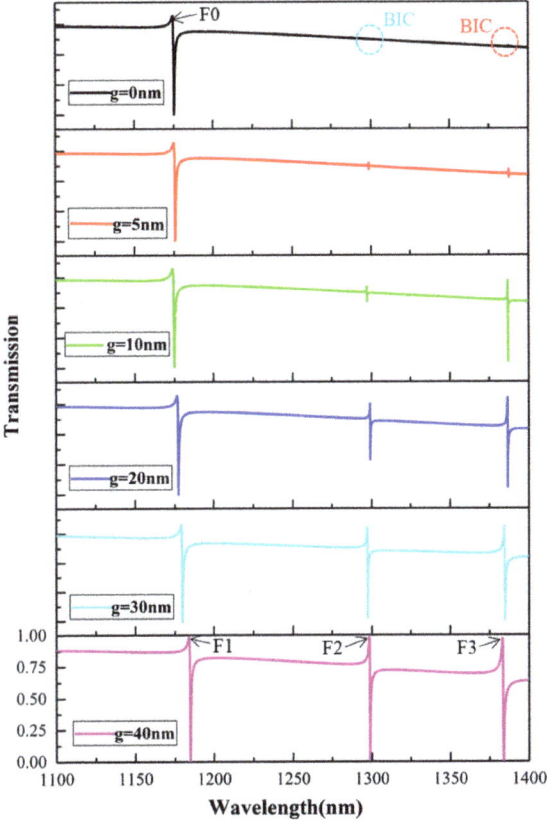

Figure 2. Transmission spectra of the SHND nanostructure with different asymmetry parameters.

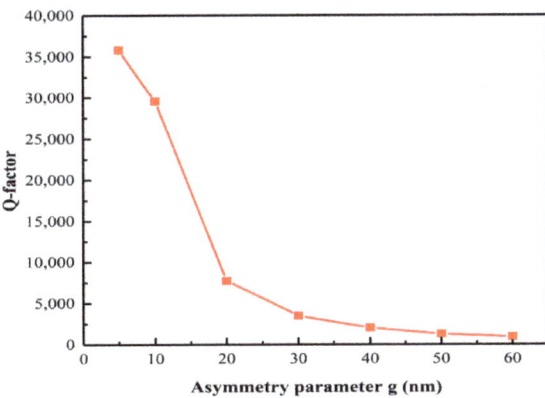

Figure 3. Q-factor of F3 with different asymmetric parameters.

In practical applications, errors in the Q-factor may be caused by factors such as technology and the environment. For example, the actual all-dielectric metasurface is uneven, which will lead to an increase in transmission loss and decrease in the Q-factor. As the substrate in our simulation is transparent glass, loss of silicon is the main consideration. By adding an extinction coefficient $k$ (i.e., the imaginary part of the refractive index of silicon) [47], the results can include the absorption and scattering losses caused by surface roughness in the real manufacturing process. As shown in Figure 4a, when $k = 10^{-11}$ or $10^{-7}$, the Q-factors decrease as g increases; when $k = 10^{-3}$, the Q-factor maintains nearly the same value. Figure 4b shows the F3 resonance curves with different $k$ values of g = 40 nm. It can be seen that when $k = 10^{-1}$, there is no transmission peak in the F3 wavelength range; when $k$ is less than $10^{-1}$, the transmission peak gradually appears in the F3 wavelength range, and the Q-factor increases accordingly. This is because decrease in the $k$ value leads to decrease in loss.

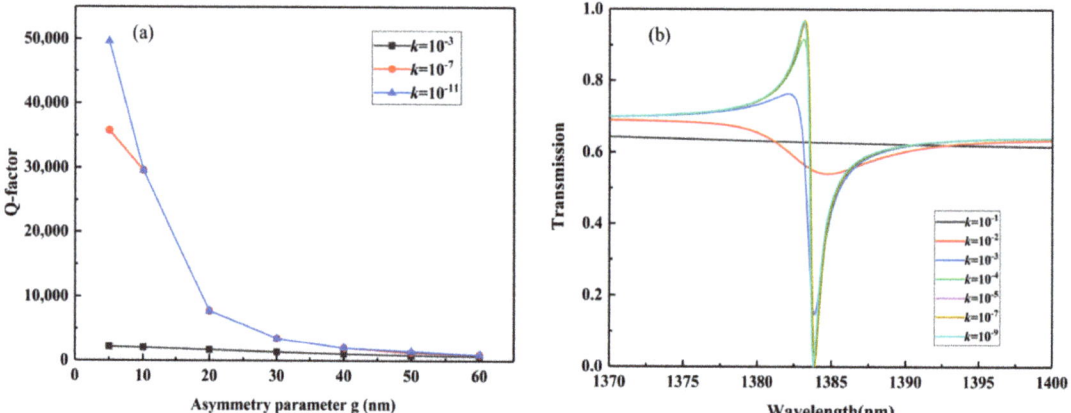

**Figure 4.** (a) Q-factor of F3 at $k = 10^{-3}$, $10^{-7}$ and $10^{-11}$ with different asymmetric parameters. (b) Transmission spectra of F3 with different extinction coefficients $k$ at g = 40 nm.

*4.2. Multipole Decomposition of Fano and Analysis of Its Resonance Mode*

To gain insight into the optical properties of the designed multi-Fano resonance device and to identify the contribution of different multipole resonance modes to the Fano resonances, the electric field and magnetic field distributions were calculated.

To understand the specific characteristics of the Fano resonance with g = 0 nm, the electric and magnetic field distributions at the resonance peak of 1172.6 nm are shown in Figure 5a,b. From the electric field distribution in Figure 5a, it can be observed that the SHND structure generates four magnetic loop currents on the x–y plane simultaneously. The MD moments of the two magnetic loop currents on the left are along the negative direction of the z-axis, whereas the MD moments of the two magnetic current loops on the right are along the positive direction of the z-axis. The TD resonance is generated, and its direction points to the negative direction of the y-axis, which is shown by the blue arrows in Figure 5a,b. This result coincides with the radiant energy distribution of each multipole resonance mode of the SHND structure with g = 0 nm, as shown in Figure 5c, indicating that the TD mode is dominant at F0.

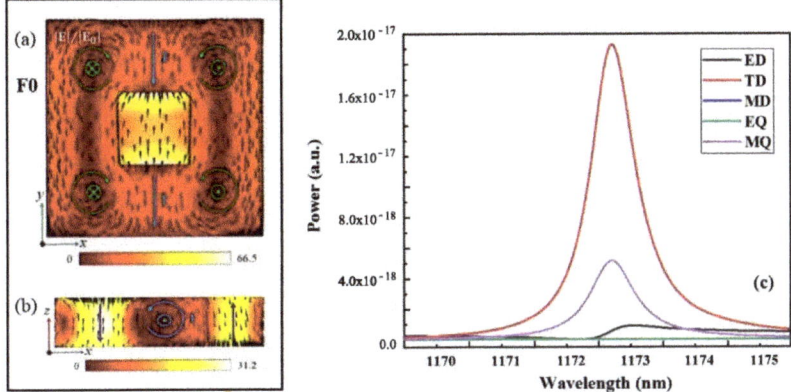

**Figure 5.** When g = 40 nm, the SHND nanostructure is symmetric. (**a**) Electric field enhancement and electric field vector distribution of F0 in the x–y plane. (**b**) Magnetic field enhancement and magnetic field vector distribution of F0 in the x z plane. (**c**) Radiant energy of each multipole resonance mode at F0.

However, further analysis shows that the amplitudes of the current Cartesian ED moment |Py| and the TD moment |ikTy| are nearly equal at 1172.6 nm, as shown in Figure 6a. Meanwhile, the phase difference of the ED and TD moments |φ(ikTy) − φ(Py)| is approximately equal to π, as shown in Figure 6b. The far-field radiation of the ED and TD resonances interfere destructively with each other; thus, the subradiant resonance anapole mode can be induced. In addition to the destructive interference between ED and TD resonances, the incident field can be efficiently trapped into the unit cell of the SHND metasurface, which plays an important role in the shaping of the resonance line type.

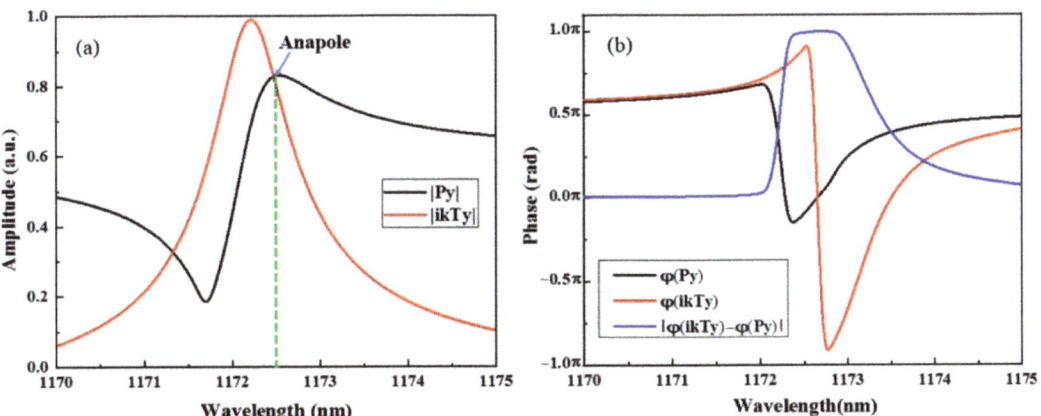

**Figure 6.** (**a**) Amplitude of the Cartesian electric dipole moment |Py| and toroidal dipole moment |ikTy| of F0 when g = 0 nm. (**b**) Phases, with the differences of Py and ikTy.

Figure 7 shows the electromagnetic field distributions of the three Fano resonances of the SHND structure with g = 40 nm. Figure 7a,b show that the electromagnetic field enhancement of F1 is greater than that of F0 and the electromagnetic field distribution of F1 changes slightly. Figure 7e shows the electric field distribution of F2 at 1294.9 nm in the x–y plane; there are also four MD resonances generated by the four magnetic loop currents in the middle of the four outer edges of the square cavity. From the combination of the magnetic field distribution in the $x$–$z$ and $y$–$z$ planes shown in Figure 7f,g, it can be inferred that two of the four magnetic loop currents in Figure 7e can be combined with each other to form TD resonance modes. Thereafter, these TD resonance modes are formed in the diagonal direction of the SHND structure, and the resonance moment directions of the two TD resonance modes on the same diagonal line are opposite. This resonance mode can be regarded as magnetic quadrupole (MQ) resonance. The multipole decomposition of F2 is shown in Figure 8a, where the energy of the magnetic quadrupole (MQ) moment is the largest, indicating that the generation of FR2 is mainly due to MQ resonance. Figure 7c,d show the electric field distribution and magnetic field distribution of F3 in the $x$–$y$ plane and $x$–$z$ plane, respectively. It can be seen from the figure that the electric field in the $x$–$y$ plane forms a ring current, and that there is a negative direction of the magnetic field along the $z$-axis. The multipole decomposition of F3 is shown in Figure 8b. It can be seen that the energy of the MD moment is the largest, so F3 is dominated by MD resonance, and the MD moment is along the negative direction of the $z$-axis.

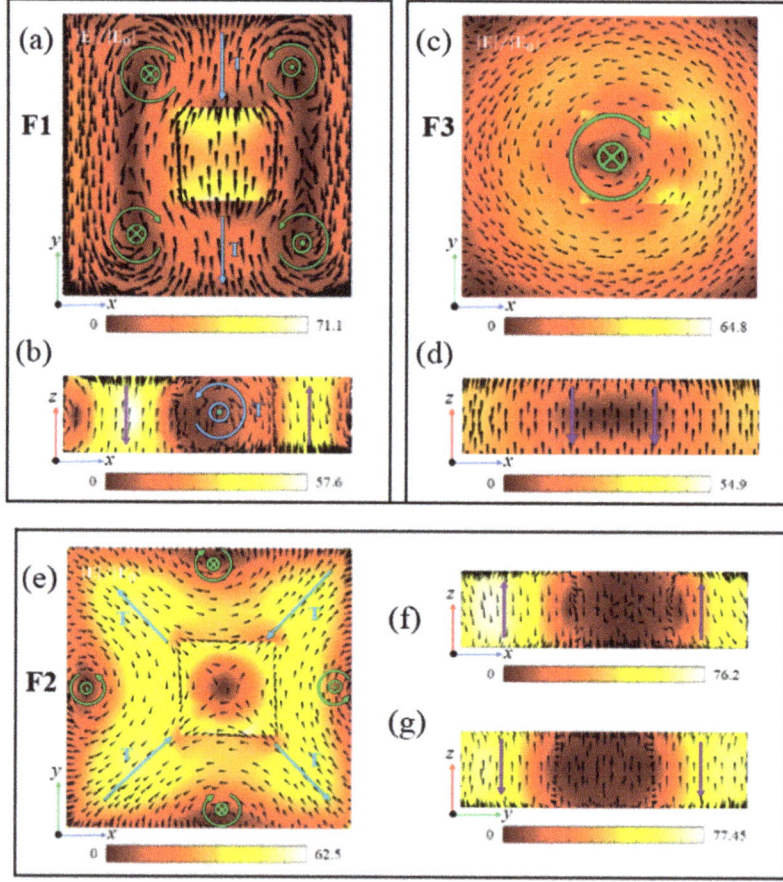

**Figure 7.** When g = 40 nm, the symmetry of the SHND nanostructure is broken. (**a,c,e**) show the electric field enhancement and electric field vector distribution of F1, F2, and F3 in the $x$–$y$ plane. (**b,d,f**) show the magnetic field enhancement and magnetic field vector distribution of F1, F2, and F3 in the $x$–$z$ plane. (**g**) shows the magnetic field enhancement and magnetic field vector distribution of F2 in the $y$–$z$ plane.

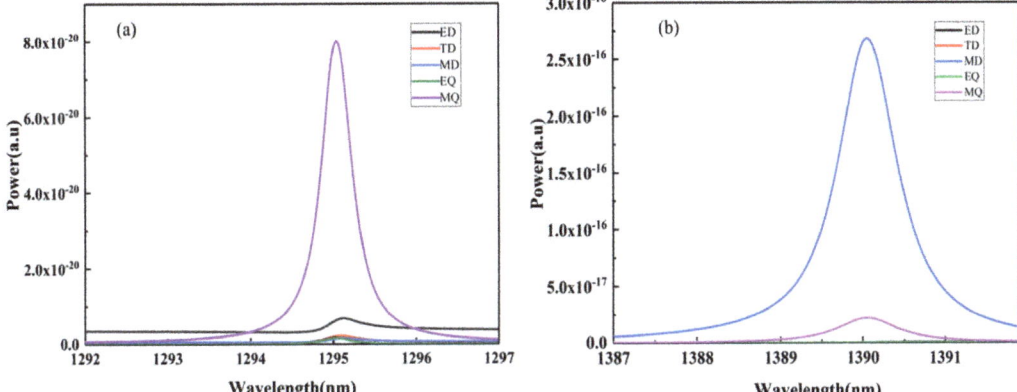

**Figure 8.** Radiant energy of each multipole resonance mode at (**a**) F2, and (**b**) F3.

By continuing to analyze the electromagnetic field enhancement features of F1, F2, and F3, it can be determined that the electromagnetic field enhancements of the three Fano resonances are nearly three times stronger than the electromagnetic field enhancement of the golden cross-shaped dimer structure [48]. In particular, the maximum electric field enhancements of F1 and F2 are found to be 71.1- and 77.45-fold, respectively; therefore, the SHND nanostructure can be applied in areas where significant near-field enhancement is required.

In addition, the modulation depth [49,50] is an effective parameter of Fano resonance, which can be used to describe the intensity range of Fano resonance. The modulation depth is usually defined as $\Delta T = T_{peak} - T_{dip}$, where $T_{peak}$ and $T_{dip}$ are the intensities of the resonance peak and dip, respectively. According to this definition, the modulation depths of F0, F1, F2, and F3 of the SHND structure at g = 0 nm and g = 40 nm are all reached at nearly 100%, which provides a technique for obtaining equipment with high modulation depth.

### 4.3. Influence of Geometric Parameters on F1, F2, and F3 of the SHND Structure

To study the dependence of the transmission spectra characteristics on the different geometric parameters of the SHND structure, the corresponding transmission spectra characteristics to variable parameters when g = 40 nm were calculated and are shown in Figure 9.

As the asymmetry parameter g value increases, it can be observed from Figure 9a that the transmission spectra of F1 undergoes a significant red shift and F3 exhibits a significant blue shift. Compared with F1 and F3, the position of F2 can be considered as not moving, which shows that F2 is not sensitive to changes in g. In addition, with increase in g, the resonance modes F1, F2, and F3 are all widened; in Figure 9b, with decrease in the outer length L, the three Fano resonances are blue-shifted. The reason is that with decrease in L, the surface area of SHND also decreases, resulting in a decrease in the effective refractive index. In Figure 9c, with increase in the SHND structure height H, the three Fano resonances show a significant simultaneous red-shift, because the increase in H leads to an increase in the effective refractive index of the SHND surface. At the same time, an increase in H has little effect on the resonant line width and Q-factor. It can also be observed from Figure 9d that, with increase in the width of the inner cavity W, the three Fano resonances also exhibit a clear blue-shift. The reason is that increase in W leads to a decrease in the SHND surface area, which leads to a decrease in the effective refractive index of the surface.

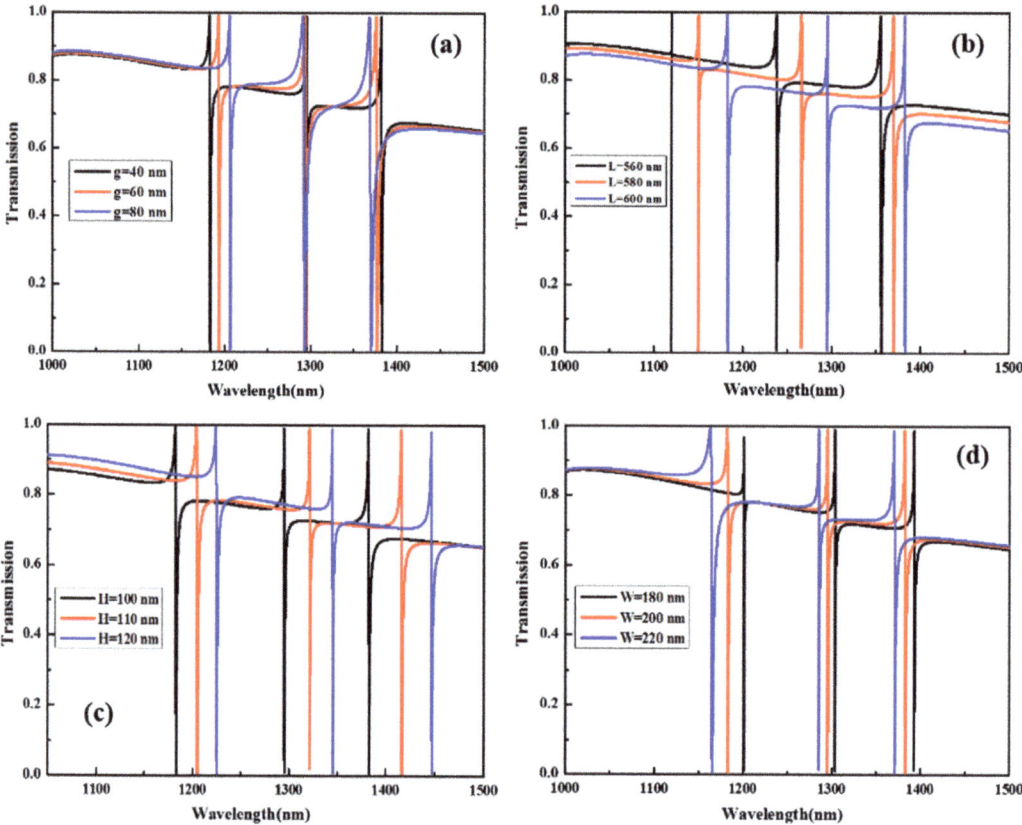

**Figure 9.** Transmission spectra of multi-Fano resonance of the SHND structure with different (**a**) asymmetry parameter g, (**b**) structure height H, (**c**) outer edge length L, and (**d**) inner cavity width W.

By comprehensively comparing and analyzing the three Fano resonance characteristics in the four cases, we can confirm that F1 is very sensitive to changes in the outer edge length L, the structural height H, and the internal cavity width W. F2 is insensitive to changes in the asymmetry parameter g, and sensitive to changes in the width of the inner cavity W, the outer length L, and the structure height H. F3 is most sensitive to changes in the structure height H. The above results show that, in practical applications, the multi-Fano resonance of the SHND metasurface can be adjusted flexibly according to the specific conditions. In addition, when g, L, H, and W are all changed in different ways, the modulation depths of the three Fano resonances F1, F2, and F3 generated by the SHND structure are all practically 100% and their FWHMs are very narrow. This result supports the design of optical devices with multiple Fano resonance, a high Q-factor, and high modulation depth.

## 5. Conclusions

We have demonstrated that, by breaking the symmetry of all-dielectric SHND metasurfaces and converting the BIC mode into a quasi-BIC mode, multi-Fano resonance with a high Q-factor and nearly 100% modulation depth can be obtained. The influence of the extinction coefficient $k$ on the Q-factor was also examined. Due to the formation of subradiation hybrid resonance modes, such as anapole and TD resonance modes, the light energy efficiency of the SHND metasurface interacting with incident light is increased, resulting in Fano resonances producing extremely narrow FHMWs and higher Q-factors. Moreover, the maximum Q-factor was found to reach 35,837 in magnitude. In addition, by changing the geometric parameters of the SHND, a larger Q-factor and a wide range of adjustments of the Fano resonance positions can be obtained. Therefore, this type of multi-Fano resonance with a high Q-factor and non-localized characteristics can enhance adaptability, enabling the SHND structure to be applied to optoelectronics and nonlinear optical devices in the visible light range, achieving higher-efficiency large-scale optoelectronic integration.

**Author Contributions:** C.Z. (Chen Zhou). and Y.W.: conceptualization, data curation, methodology, software, and writing–original draft; Y.H.: review and editing, supervision, project administration, funding acquisition; P.C., M.S., T.L., C.Z. (Chen Zhao). and Z.L.: review and editing; Z.Z. and Y.X.: supervision. All authors have read and agreed to the published version of the manuscript.

**Funding:** The National Natural Foundation of China (Grant No. 11604198).

**Institutional Review Board Statement:** Not applicable.

**Informed Consent Statement:** Not applicable.

**Data Availability Statement:** The data presented in this study are available on request from the corresponding author.

**Conflicts of Interest:** The authors declare no conflict of interest.

## References

1. Yang, Y.; Wang, W.; Boulesbaa, A.; Kravchenko, I.I.; Briggs, D.P.; Puretzky, A.; Geohegan, D.; Valentine, J. Nonlinear Fano-Resonant Dielectric Metasurfaces. *Nano Lett.* **2015**, *15*, 7388–7393. [CrossRef] [PubMed]
2. Cesca, T.; Manca, M.; Michieli, N.; Mattei, G. Tuning the linear and nonlinear optical properties of ordered plasmonic nanoarrays by morphological control with thermal annealing. *Appl. Surf. Sci.* **2019**, *491*, 67–74. [CrossRef]
3. Zou, C.J.; Sautter, J.; Setzpfandt, F.; Staud, I. Resonant dielectric metasurfaces: Active tuning and nonlinear effects. *J. Phys. D Appl. Phys.* **2019**, *52*, 373002. [CrossRef]
4. Baghbadorani, H.K.; Barvestani, J. Sensing improvement of 1D photonic crystal sensors by hybridization of defect and Bloch surface modes. *Appl. Surf. Sci.* **2020**, *537*, 147730. [CrossRef]
5. Cho, C.H.; Aspetti, C.O.; Park, J.; Agarwal, R. Silicon coupled with plasmon nanocavities generates bright visible hot luminescence. *Nat. Photonics* **2013**, *4*, 285–289. [CrossRef]
6. Zhou, Y.J.; Dai, L.H.; Li, Q.Y.; Xiao, Z.Y. Two-Way Fano Resonance Switch in Plasmonic Metamaterials. *Front. Phys.* **2020**, *8*, 576419. [CrossRef]
7. Campione, S.; Liu, S.; Basilio, L.I.; Warne, L.K.; Langston, W.L.; Luk, T.S.; Wendt, J.R.; Reno, J.L.; Keeler, G.A.; Brener, I.; et al. Broken symmetry dielectric resonators for high quality-factor Fano metasurfaces. *ACS Photonics* **2016**, *3*, 2362–2367. [CrossRef]
8. Moritake, Y.; Kanamori, Y.; Hane, K. Enhanced quality factor of Fano resonance in optical metamaterials by manipulating configuration of unit cells. *Appl. Phys. Lett.* **2015**, *21*, 211108. [CrossRef]
9. Zhang, T.; Wang, J.; Liu, Q.; Zhou, J.; Dai, J.; Han, X.; Li, J.; Zhou, Y.; Xu, K. Efficient spectrum prediction and inverse design for plasmonic waveguide systems based on artificial neural networks. *Photonics Res.* **2019**, *3*, 368. [CrossRef]
10. He, F.; Wang, M.; Jiao, L.; Xu, Z.; Yun, M. Phase-Coupled Plasmon-Induced Transparency in metasurface with Periodically Arranged Bimolecular Systems. *Appl. Surf. Sci.* **2020**, *506*, 144888.
11. Mohsen, R.; Lei, D.Y.; Giannini, V.; Lukiyanchuk, B.; Ranjbar, M.; Liew, T.Y.F.; Hong, M.; Maier, S.A. Subgroup Decomposition of Plasmonic Resonances in Hybrid Oligomers: Modeling the Resonance Line shape. *Nano Lett.* **2012**, *12*, 2101–2106.
12. Zhao, C.; Cui, L.; Song, X.; Xiao, J. High sensitivity plasmonic sensing based on Fano interference in a rectangular ring waveguide. *Opt. Commun.* **2015**, *6*, 1817–1824.
13. Li, Y.; Huo, Y.; Zhang, Y.; Zhang, Z. Generation and Manipulation of Multiple Magnetic Fano Resonances in Split Ring-Perfect/Ring Nanostructure. *Plasmonics* **2016**, *5*, 1613–1619. [CrossRef]
14. Zhang, Q.; Wen, X.; Li, G.; Ruan, Q.; Wang, J.; Xiong, Q. Multiple Magnetic Mode-Based Fano Resonance in Split-Ring Resonator/Disk Nanocavities. *ACS Nano* **2013**, *12*, 11071–11078. [CrossRef] [PubMed]
15. Yang, Z.J. Fano Interference of Electromagnetic Modes in Subwavelength Dielectric Nanocrosses. *J. Phys. Chem. C* **2016**, *38*, 21843–21849. [CrossRef]
16. Liu, H.; Zheng, L.; Ma, P.; Zhong, Y.; Liu, B.; Chen, X.; Liu, H. Metasurface generated polarization insensitive Fano resonance for high-performance refractive index sensing. *Opt. Express* **2019**, *9*, 13252–13267. [CrossRef]
17. Hsu, C.W.; Zhen, B.; Stone, A.D.; Joannopoulos, J.D.; Soljačić, M. Bound states in the continuum. *Nat. Rev. Mater.* **2016**, *1*, 16048. [CrossRef]
18. Zhou, C.B.; Pu, T.Y.; Huang, J.; Fan, M.H.; Huang, L.J. Manipulating Optical Scattering of Quasi-BIC in Dielectric Metasurface with Off-Center Hole. *Nanomaterials* **2022**, *12*, 54. [CrossRef]

19. Tian, J.Y.; Li, Q.; Belov, P.A.; Sinha, R.K.; Qian, W.P.; Qiu, M. High-Q All-Dielectric Metasurface: Super and Suppressed Optical Absorption. *ACS Photonics* **2022**, *7*, 1436–1443. [CrossRef]
20. Huo, Y.Y.; Zhang, X.; Yan, M.; Sun, K.; Jiang, S.Z.; Ning, T.Y.; Zhao, L.N. Highly-sensitive sensor based on toroidal dipole governed by bound state in the continuum in dielectric non-coaxial core-shell cylinder. *Opt. Express* **2022**, *30*, 19030–19041. [CrossRef]
21. Benea-Chelmus, I.-C.; Mason, S.; Meretska, M.L.; Elder, D.L.; Kazakov, D.; Shams-Ansari, A.; Dalton, L.R.; Capasso, F. Gigahertz free-space electro-optic modulators based on Mie resonances. *Nat. Commun.* **2022**, *13*, 3170. [CrossRef] [PubMed]
22. Evlyukhin, A.B.; Fischer, T.; Reinhardt, C.; Chichkov, B.N. Optical theorem and multipole scattering of light by arbitrarily shaped nanoparticles. *Phys. Rev. B* **2016**, *20*, 205434. [CrossRef]
23. Hu, P.; Liang, L.; Ge, L.; Xiang, H.; Han, D. Fano resonance induced by the toroidal moment in cylindrical metallic meta-structures. *Optics* **2019**, *21*, 055001. [CrossRef]
24. Shi, Y.; Zhou, L.-M.; Liu, A.Q.; Nieto-Vesperinas, M.; Zhu, T.; Hassanfiroozi, A.; Liu, J.; Zhang, H.; Tsai, D.P.; Li, H.; et al. Superhybrid Mode-Enhanced Optical Torques on Mie-Resonant Particles. *Nano Lett.* **2022**, *22*, 1769–1777. [CrossRef]
25. Zel'Dovich, I.B. Electromagnetic Interaction with Parity Violation. *J. Exp. Theor. Phys.* **1958**, *6*, 1184.
26. Kaelberer, T.; Fedotov, V.A.; Papasimakis, N.; Tsai, D.P. Toroidal dipolar response in a metamaterial. *Science* **2010**, *6010*, 1510–1512. [CrossRef]
27. Hopkins, B.; Filonov, D.S.; MiroSHNDichenko, A.E.; Monticone, F.; Alù, A.; Kivshar, Y.S. Interplay of Magnetic Responses in All-Dielectric Oligomers to Realize Magnetic Fano Resonances. *ACS Photonics* **2015**, *2*, 724–729. [CrossRef]
28. Kroychuk, M.K.; Shorokhov, A.S.; Yagudin, D.F.; Shilkin, D.A.; Smirnova, D.A.; Volkovskaya, I.; Shcherbakov, M.R.; Shvets, G.; Fedyanin, A.A. Enhanced Nonlinear Light Generation in Oligomers of Silicon Nanoparticles under Vector Beam Illumination. *Nano Lett.* **2020**, *20*, 3471–3477. [CrossRef]
29. Liu, Z.; Li, J.; Li, W.; Li, J.; Gu, C.; Li, Z. 3D conductive coupling for efficient generation of prominent Fano resonances in metamaterials. *Sci. Rep.* **2016**, *6*, 27817. [CrossRef]
30. Luk'Yanchuk, B.; Zheludev, N.I.; Maier, S.A.; Halas, N.J.; Nordlander, P.; Giessen, H.; Chong, C.T. The Fano resonance in plasmonic nanostructures and metamaterials. *Nat. Mater.* **2010**, *9*, 707–715. [CrossRef]
31. Miroshnichenko, A.E.; Evlyukhin, A.B.; Yu, Y.F.; Bakker, R.M.; Chipouline, A.; Kuznetsov, A.; Luk'Yanchuk, B.; Chichkov, B.N.; Kivshar, Y.S. Nonradiating anapole modes in dielectric nanoparticles. *Nat. Commun.* **2015**, *6*, 8069. [CrossRef] [PubMed]
32. Papasimakis, N.; Fedotov, V.A.; Savinov, V.; Raybould, T.A.; Zheludev, N. Electromagnetic toroidal excitations in matter and free space. *Nat. Mater.* **2016**, *3*, 263–271. [CrossRef] [PubMed]
33. Walsh, G.F.; Negro, L.D. Enhanced Second Harmonic Generation by Photonic–Plasmonic Fano-Type Coupling in Nanoplasmonic Arrays. *Nano Lett.* **2016**, *7*, 3111–3117. [CrossRef] [PubMed]
34. Liu, S.; Wang, Z.; Wang, W.; Chen, J.; Chen, Z. High Q-factor with the excitation of anapole modes in dielectric split nanodisk arrays. *Opt. Express* **2017**, *19*, 22375–22387. [CrossRef]
35. Sun, G.; Yuan, L.; Zhang, Y.; Zhang, X.; Zhu, Y. Q-factor enhancement of Fano resonance in all-dielectric metasurfaces by modulating meta-atom interactions. *Sci. Rep.* **2017**, *1*, 8128. [CrossRef]
36. Tao, Y.; Guo, Z. Molecular detection by active Fano-sensor. *Ann. Phys.* **2017**, *4*, 1600259.
37. Tian, X.; Fang, Y.; Zhang, B. Multipolar Fano Resonances and Fano-Assisted Optical Activity in Silver Nanorice Heterodimers. *ACS Photonics* **2014**, *11*, 1156–1164. [CrossRef]
38. Algorri, J.F.; Zografopoulos, D.C.; Sanchez-Pena, J.M. Ultrahigh-quality factor resonant dielectric metasurfaces based on hollow nanocuboids. *Opt. Express* **2019**, *27*, 6320–6330. [CrossRef]
39. Algorri, J.F.; Zografopoulos, D.C.; Ferraro, A.; Garcia, C.; Vergaz, B.; Beccherelli, R.; Sanchez, P.; Jose, M. Anapole Modes in Hollow Nanocuboid Dielectric Metasurfaces for Refractometric Sensing. *Nanomaterials* **2019**, *9*, 30.
40. Jeong, J.; Goldflam, M.D.; Campione, S.; Briscoe, J.L.; Vabishchevich, P.P.; Nogan, J.; Sinclair, M.B.; Luk, T.S.; Brener, I. High Quality Factor Toroidal Resonances in Dielectric Metasurfaces. *ACS Photonics* **2020**, *7*, 1699–1707. [CrossRef]
41. Palik, E.D. *Handbook of Optical Constants of Solids*; Academic Press: Cambridge, MA, USA, 1985; Volume I.
42. Tasolamprou, A.C.; Tsilipakos, O.; Kafesaki, M.; Soukoulis, C.M.; Economou, E.N. Toroidal eigenmodes in all-dielectric metamolecules. *Phys. Rev. B* **2016**, *20*, 205433. [CrossRef]
43. Yang, Y.; Zenin, V.A.; Bozhevolnyi, S.I. Anapole-Assisted Strong Field Enhancement in Individual All-Dielectric Nanostructures. *ACS Photonics* **2018**, *5*, 1960–1966. [CrossRef]
44. Lamprianidis, A.G.; Miroshnichenko, A.E. Excitation of nonradiating magnetic anapole states with azimuthally polarized vector beams. *Beilstein J. Nanotechol.* **2018**, *9*, 1478–1490. [CrossRef] [PubMed]
45. Yang, Y.; Kravchenko, I.I.; Briggs, D.P.; Valentine, J. All-dielectric metasurface analogue of electromagnetically induced transparency. *Nat. Commun.* **2014**, *5*, 5753. [CrossRef] [PubMed]
46. Yang, L.; Yu, S.L.; Li, H.; Zhao, T.G. Multiple Fano resonances excitation on all-dielectric nanohole arrays metasurfaces. *Opt. Express* **2021**, *29*, 14905–14916. [CrossRef]
47. Algorri, J.F.; Dell'Olio, F.; Roldan-Varona, P.; Rodriguez-Cobo, L.; Lopez-Higuera, J.M.; Sanchez-Pena, J.M.; Zografopoulos, D.C. Strongly resonant silicon slot metasurfaces with symmetry-protected bound states in the continuum. *Opt. Express* **2021**, *29*, 10374–10385. [CrossRef]
48. Zhang, J.; Chen, S.; Wang, J.; Mu, K.; Fan, C.; Liang, E.; Pei, D. An engineered CARS substrate with giant field enhancement in crisscross dimer nanostructure. *Sci. Rep.* **2018**, *1*, 740. [CrossRef]

49. Zhang, Y.; Liu, W.; Li, Z.; Li, Z.; Cheng, H.; Chen, S.; Tian, J. High-quality-factor multiple Fano resonances for refractive index sensing. *Opt. Lett.* **2019**, *2*, 2818–2825. [CrossRef]
50. Zhao, W.; Jiang, H.; Liu, B.; Jiang, Y.; Tang, C.; Li, J. Fano resonance based optical modulator reaching 85% modulation depth. *Appl. Phys. Lett.* **2015**, *17*, 171109. [CrossRef]

*Article*

# Efficient Achromatic Broadband Focusing and Polarization Manipulation of a Novel Designed Multifunctional Metasurface Zone Plate

Shaobo Ge, Weiguo Liu *, Xueping Sun, Jin Zhang, Pengfei Yang *, Yingxue Xi, Shun Zhou, Yechuan Zhu and Xinxin Pu

Shaanxi Province Key Laboratory of Thin Films Technology and Optical Test, School of Optoelectronic Engineering, Xi'an Technological University, Xi'an 710032, China; geshaobo@126.com (S.G.); xuepingsun@xatu.edu.cn (X.S.); j.zhang@xatu.edu.cn (J.Z.); xiyingxue@163.com (Y.X.); zsemail@126.com (S.Z.); zyc_xatu@126.com (Y.Z.); pxx_1125@126.com (X.P.)
* Correspondence: wgliu@163.com (W.L.); pfyang@xatu.edu.cn (P.Y.); Tel.: +86-029-83208114 (W.L.)

**Abstract:** In this paper, comprehensively utilizing the diffraction theory and electromagnetic resonance effect is creatively employed to design a multifunctional metasurface zone plate (MMZP) and achieve the control of polarization states, while maintaining a broadband achromatic converging property in a near-IR region. The MMZP consists of several rings with fixed width and varying heights; each ring has a number of nanofins (usually called meta-atoms). The numerical simulation method is used to analyze the intensity distribution and polarization state of the emergent light, and the results show that the designed MMZP can realize the polarization manipulation while keeping the broadband in focus. For a specific design wavelength (0.7 μm), the incident light can be converted from left circularly polarized light to right circularly polarized light after passing through the MMZP, and the focusing efficiency reaches above 35%, which is more than twice as much as reported in the literature. Moreover, the achromatic broadband focusing property of the MMZP is independent with the polarization state of the incident light. This approach broadens degrees of freedom in micro-nano optical design, and is expected to find applications in multifunctional focusing devices and polarization imaging.

**Keywords:** metasurface; zone plate; achromatic

## 1. Introduction

To make optical components lightweight and multifunctional has always been one of the goals pursued in optics. Fresnel zone plate (FZP) is a typical representative [1–8]. In particular, developed on the basis of FZP, the multi-level diffraction lens (MDL) has achieved an achromatic imaging function range from visible to long-wave infrared bands via a globally optimized numerical iterative algorithm [9,10]. However, it can not modulate the polarization state of incident light while focusing with wide spectrum achromatic [11–13].

It is worth noting that the proposal of metasurface opens a new field of vision in the multi-functional design of optical components [14–17]. The free manipulation of the amplitude, phase and polarization of light by the metasurface has completely broken the limitation of optical materials [18,19]. Recently, many research studies devoted to the polarization transformation based on metasurface have appeared. Some of these works are also considered multifunctional optical elements which provide polarization transformation and focusing simultaneously [20–23].

Gwanho Yoon et al. proposed a new type metasurface called metasurface zone plates (MZP), achieved focusing and polarization manipulation for a single-wavelength via replacing the typical FZP rings by metasurface [24]. At present, MZP can achieve single-wavelength polarization conversion while focusing on several discrete wavelengths by using metal or dielectric subwavelength nano-antennas [25]. In addition, the focusing efficiency of these MZPs is generally low, just around 10% [26–30], and the highest focusing efficiency reported so far is 17% [31]. Obviously, it is still a very challenging task to achieve efficient broadband achromatic focusing while ensuring the control of polarization states. In my opinion, the main reason why MZP is not efficient is that all these research works replace the rings via metasurface, which discards the powerful amplitude control of traditional diffraction elements.

In this research, a novel nested composite structured multifunctional metasurface zone plate (MMZP) is designed via the combination of the diffraction theory and the electromagnetic resonance effect, which formed by integrating metasurface on the surface of the MDL rings. Based on the global optimization mathematical iterative method, the height distribution of the MMZP is optimized to realize the highly efficient achromatic broadband focus. Furthermore, the polarization state of incident light is accurately regulated by scanning and iterating the dimension parameters of the composite structure. This combination broadens degrees of freedom in micro-nano optical design, and is expected to find applications in multifunctional focusing devices, polarization imaging, and other fields [32–34].

## 2. Methods

The MMZP consists of several rings with fixed width and varying heights, each with a number of nanofins (usually called meta-atoms) above it. The high focusing efficiency over broadband wavelengths is achieved by selecting the multiple height levels dictated by the nonlinear optimization methodology. The polarization state of light can be manipulated, benefitting from the advantage of extreme form birefringence of metasurface. Herein, employing the diffraction theory and strong electromagnetic resonances simultaneously, broadband focusing and polarization-modulation are achieved. A left circularly polarized (LCP) light can be trasformed as a right circularly polarized (RCP) light. Figure 1a shows a schematic diagram for the MMZP.

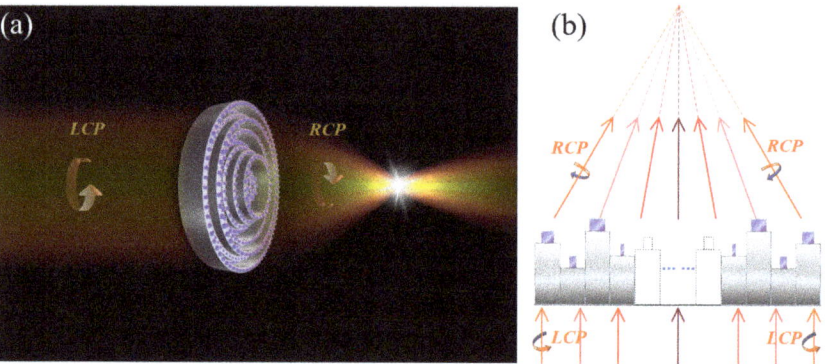

**Figure 1.** Schematic diagram of the MMZP designed for broadband focusing and polarization manipulation. The colors of the beams represent the different wavelengths. The arrow represents the spin state of the beam. (**a**) Three-dimensional diagram. (**b**) Sectional view.

To design an MMZP, the multi-level diffraction lens should be modeled first. After that, the phase profiles of each ring can be generated by changing the geometry parameters (length and width) of the nanofins. Then, the appropriate electromagnetic response produces the desired optical characteristics by the combination of MDL and metasurface as a novel hybrid design.

### 2.1. Design of Broadband MDL

The MDL design aims at high focusing efficiency over all wavelengths interested using a direct binary search (DBS) algorithm [35]. In essence, the design of MDL is a process of inversely solving the height distribution on the premise of knowing the desired optical field distribution of the imaging plane. For the broadband wavelengths, the diffracted field at the imaging plane is given by the Fresnel transformation [36]:

$$U(x',y';\lambda) = \frac{e^{ikd}}{i\lambda d} \iint g_{illum}(x,y;\lambda) T(x,y;\lambda) e^{i\frac{k}{2d}[(x'-x)^2+(y'-y)^2]} dxdy \quad (1)$$

where $x'$ and $y'$ are the coordinates of the image plane. $x$ and $y$ are the coordinates of the MDL plane. $\lambda$ is the wavelength, and $k = \frac{2\pi}{\lambda}$ is the wave number. $d$ is the propagation distance. $g_{illum} = 1$ when the unit amplitude illumination wave is on-axis.

The corresponding transmission function of the MDL is

$$T(x,y;\lambda) = e^{i\phi(x,y;\lambda)} = e^{ik\Delta h(n-1)} \quad (2)$$

where $\Delta h = \frac{h_{max}}{N_{levels}}$ is the height perturbation; $h_{max}$ is the maximum height of the profile; $N_{levels}$ is the total number of quanitization levels. Meanwhile, the $x$ and $y$ coordinates determine the ring width of MDL [36].

When designing the broadband MDL, the weight factor (noted as $\omega_i$) is introduced into the above model. Moreover, the continuous band is separated into $N$ parts of operating wavelengths (where $N$ is the given positive integer). At this point, the maximal average focusing efficiency of the broadband can be found by adjusting (increasing or decreasing operations for example) the height of each ring, while the width of each ring is fixed.

Obviously, the design problem of MDL has been transformed into a mathematical optimization problem. The numerical iteration algorithm is implemented to solve this nonlinear optimization problem. Given an initial height distribution of MDL, positive or negative height perturbation ($\Delta h$) is applied to each groove until the iteration termination condition is satisfied. Here, the iteration stop condition is defined as the figure-of-merit (FOM), which is coupled with the average focusing efficiency. The FOM is defined by [36]

$$FOM = \frac{\sum_{i=2}^{N} \omega_i \mu_i}{N} - 10 \times \frac{\sum_{i=2}^{N} \omega_i \epsilon_i}{N} \quad (3)$$

where $\omega_i$ is the weight factor to balance contributions from different wavelength. $N$ is the total number of the wavelengths. $\mu_i$ is called the efficiency, and $\epsilon_i$ is the normalized absolute difference, which can be expressed by the following two equations [36]:

$$\mu_i = \frac{\iint I_i(x',y') F_i(x',y') dx'dy'}{\iint I_i(x',y') dx'dy'} \quad (4)$$

$$\epsilon_i = \frac{\iint |normalize(I_i(x',y')) - F_i(x',y')| dx'dy'}{\iint dx'dy'} \quad (5)$$

Here, $I_i(x',y') = |U(x',y';\lambda)|^2$ is the intensity at the image plane of the $i$-th wavelength. As the first-order approximation of a focusing point-spread-function (PSF), the objective function ($F_i(x',y')$) is defined as a Gaussian function centered at $\left(\frac{x'_{min}+x'_{max}}{2}, \frac{y'_{min}+y'_{max}}{2}\right)$ with full-width-at-half-maximum (FWHM) $W_i$ determined by the far-field diffraction limit [36,37]:

$$F_i(x',y') = \exp\left\{-\frac{\left(x' - \frac{x'_{min}+x'_{max}}{2}\right)^2 + \left(y' - \frac{y'_{min}+y'_{max}}{2}\right)^2}{\left(\frac{W_i}{2}\right)^2}\right\} \quad (6)$$

$$W_i = \frac{\lambda_i}{2NA} \quad (7)$$

$$NA = \sin\left[\tan^{-1}\left(\frac{D/2}{f}\right)\right] \tag{8}$$

where $x'_{min}$, $x'_{max}$, $y'_{min}$ and $y'_{max}$ delimit the integration range from the leftmost to the rightmost of the MDL design. $\lambda_i$ is the $i$-th incident wavelength. $NA$ is the numerical aperture. $D$ is the diameter of the MDL and $f$ is the designed focal length.

Note that the efficiency $\mu_i$ is defined proportionally to the focusing efficiency $\eta$, which is used as the power ratio of the focal spot (with a radius of just three times the FWHM spot size) to the total incident optical power. This means that the termination of the iterative calculation is conditional on maximizing the focusing efficiency. When the expect parameters (focal length, element aperture, ring width, and the material refractive index) are determined, the optimized height of each ring can be evaluated.

## 2.2. Design of the MMZP

To realize the polarization manipulation while maintaining broadband focusing, dielectric metasurfaces are introduced for the transmission phase modulation based on the MDL design. Metasurface is often used to generate new optical elements according to the geometric Pancharatnam-Berry (P-B) phase and dynamic phase. The P-B phase metasurface change the orientation angle of the nanofins to realze the dependent phase of transmitted or reflected light. For the dynamic phase metasurface, the phase modulation can be chieved by changing the geometry of the nanofin. Due to the interplay of the P-B phase and dynamic phase, dielectric metasurfaces can generate arbitrary polarization states, allowing light manipulation in the vectorial regime. In addition, the physical mechanism of the broadband focusing is based on the diffraction theory. It can be seen that the polarization characteristics and broadband focusing characteristics are independent of each other in our design.

When a birefringence nanofin is illuminated by a linear polarization light, the relation between the input ($E^i$) and the output ($E^o$) electric fields can be expressed as follows [38]:

$$\begin{bmatrix} E_x^o \\ E_y^o \end{bmatrix} = R(-\theta) \begin{bmatrix} e^{i\phi_x} & 0 \\ 0 & e^{i\phi_y} \end{bmatrix} R(\theta) \begin{bmatrix} E_x^i \\ E_y^i \end{bmatrix} \tag{9}$$

where $E_x^i$ and $E_y^i$ are the input electric field components of the $x$ and $y$ direction, and $E_x^o$ and $E_y^o$ are the output electric field components of the $x$ and $y$ direction. $\theta$ is the orientation angle of the anisotropic meta-atoms. Denote $\phi_x$ and $\phi_y$ as phase delays of the meta-atoms for $x$-linearly polarized (XLP) and $y$-linearly polarized (YLP) light. As we know, a linearly polarized (LP) optical wave can be viewed as a linear superposition of a LCP and a RCP optical wave. By selecting a group of suitable parameters of the nanofin, the relation between the modulation phase and the anisotropic meta-atoms' properties is given by the formula [38]:

$$\begin{cases} |\phi_x - \phi_y| = \pi \\ \phi^+(x,y) = \phi_x + 2\theta \\ \phi^-(x,y) = \phi_y - 2\theta \end{cases} \tag{10}$$

where $\phi^+(x,y)$ and $\phi^-(x,y)$ are the modulation phase of two arbitrary orthogonal states of polarization, which are based on $\phi_x$ and $2\theta$. The geometrical size and the refractive index of the nano-fin determin the dynamic phase $\phi_x$. The PB phase $2\theta$ is related to the orientation angle.

According to this equation, when a series of meta-atoms with the phase ($\phi_x$, $\phi_y$) satisfy Equation (10), dielectric metasurfaces can achieve complete phase coverage to cover the whole Poincaré sphere in the design wavelength [24]. In our research, the LCP optical wave will be transformed into the right circularly polarization state after passing through the nanofin when the phase difference between $\phi_x$ and $\phi_y$ is $\pi$.

The phase difference created by the waveguide effect is described as $\frac{2\pi}{\lambda} n_{eff} H$, where $n_{eff}$ and $H$ are the effective index and height of the unit cell [31]. Here, $H$ is limited by the height of each ring in MDL. Then, the value of $n_{eff}$ can be changed by adjusting the geometrical size of the nanostructure when the material is identified. The phase difference can be numerically analyzed by the Finite-Difference Time-Domain (FDTD) solver (Lumerical Solutions, Inc., Vancouver, BC, Canada). Different from the classical MZP design, the ring width and ring height of MDL limit the period and height of the unit cell. Besides, the nanostructure parameters of each ring need to be optimized separately. Finally, the characteristics of broadband focusing and polarization regulation can be achieved simultaneously after combining the nano-fins on each ring, which maintains the same height as the MDL.

## 3. Results and Discussion

### 3.1. Broadband Focusing Characteristic of MDL

The MDL is designed for the wavelength range from 0.7 μm to 0.8 μm, which is a typical photoelectric detection band. The diameter is set as 9.9 μm, considering the computational load of numerical simulation. The width of each ring is fixed as 0.3 μm, which limits the period of the meta-atom for the polarization regulation in next step. Considering manufacturability, the material of the MDL is selected as $SiO_2$ thin film. $NA$ is 0.9 to get a good focusing effect. The iterative algorithm mentioned earlier is implemented programmatically. After 122 times of iterations, the optimized height distribution of the MDL is shown in Figure 2a. The maximum height is 3.5 μm, which provides a degree of freedom for phase control design. The height distribution is input into the FDTD commercial software to simulate optical field. Perfectly matched layers (PML) are applied at x and y directions and the propagation direction z.

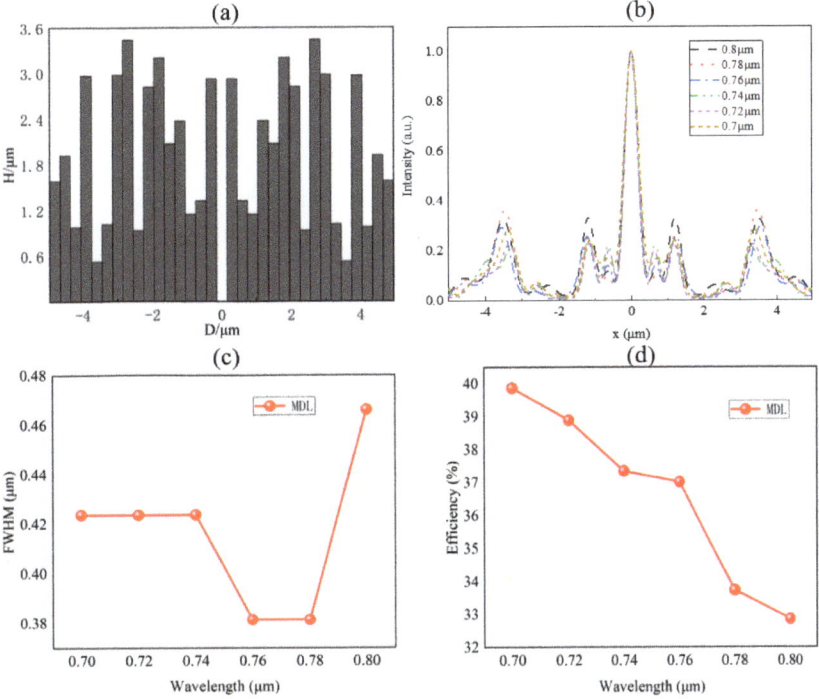

**Figure 2.** Focusing characteristic of the MDL. (**a**) Optimized height distribution. (**b**) Normalized intensity distribution along transversal axis (*x* axis). (**c**) FWHM and (**d**) Focusing efficiency at corresponding focal plane as a function of wavelength.

Figure 2b shows the intensity distributions at the x–z plane. Focusing is accompanied with a sidelobe, which is also a common phenomenon for the diffraction elements. The FWHM as function of wavelength is plotted in Figure 2c. It is noted that all wavelengths achieve sub-wavelength focusing. The focusing efficiency is shown in Figure 2d. The average focusing efficiency in the near infrared band is above 33%, which is nearly 40% at 0.7 µm wavelength.

### 3.2. Hybrid Designed Results of the MMZP

To realize the phase difference of $\pi$ between x-polarized and y-polarized light incident, the unit cell of the metasurface zone plate should be designed rigorously. The unit cell consists of a $SiN_x$ meta-atom embedded on the $SiO_2$ cell of the MDL ring as shown in Figure 3a,b. The high refractive index helps to reduce the aspect ratio of the structure. Si is not selected because the refractive index difference between upper and lower materials is too large, which will produce strong interference effect to interfere with broadband focusing [39]. The $SiO_2$ and $SiN_x$ used here have refractive index of 1.5 and 2.1, which can be prepared by plasma enhanced chemical vapor deposition (PECVD) [40].

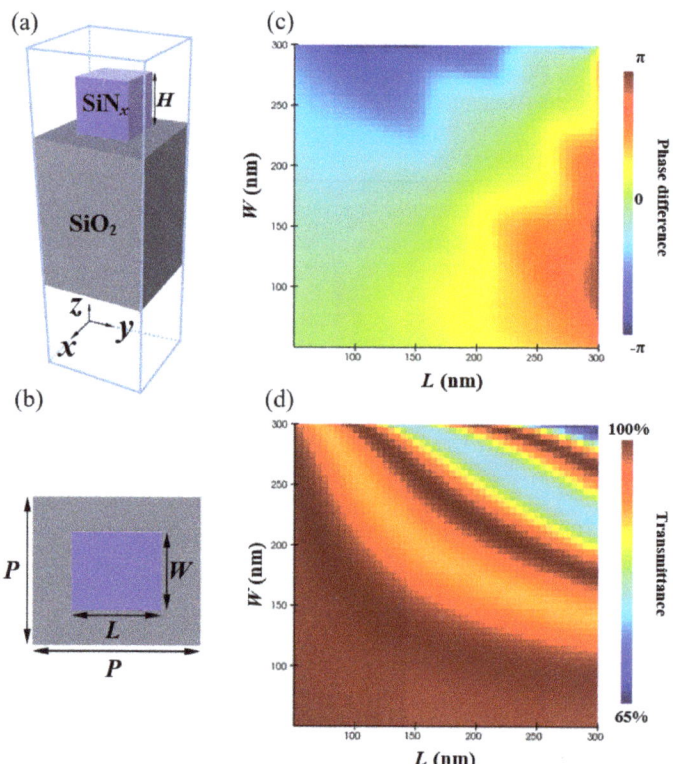

**Figure 3.** (a) Side-view and (b) top-view of the the unit cell: the period of a cell is P, and the length, width, and height are L, W, and H, respectively. (c) Simulated phase difference and (d) Simulated transmittance as a function of the nanofin size for the incident light of 0.7 µm.

The FDTD method is implemented to obtain the phase difference range and the transmission coefficients as shown in Figure 3c,d. The incident wavelength is plane wave with an x or y polarization and propagates along +z direction. Periodic boundary conditions (PBC) are applied at y directions and perfectly matched layers (PML) at the directions x and z. Here, the height of the meta-atom is limited by the height distribution of the MDL, and the refractive index of the material is also determined. The length ($L$) and width ($W$) swepting from 50 to 300 nm of the nanofin can be changed to obtain the suitable value for the phase difference of $|\phi_x - \phi_y|$. As can been see from the results, the corresponding phase difference span from $-\pi$ to $\pi$, and meanwhile the transmissions can reach up to more than 90%. For example, for the first ring, the height of $SiO_2$ is set as 0.5 µm, the height of $SiN_x$ is set as 1.1 µm. We select $L = 290$ nm and $W = 110$ nm, then the phase difference can be $\pi$, which is our expected value. This calculation should be performed 17 times to obtain the required structural parameters. Figure 4 shows the height distribution of MDL and MMZP. Both are 9.9 µm in diameter. They maintain the same height distribution. The difference is that MMZP is composed of nested composite structures. The meta-atom building blocks are arranged into periodic arrays, while the three-dimensional drawing of designed MMZP is shown in Figure 5.

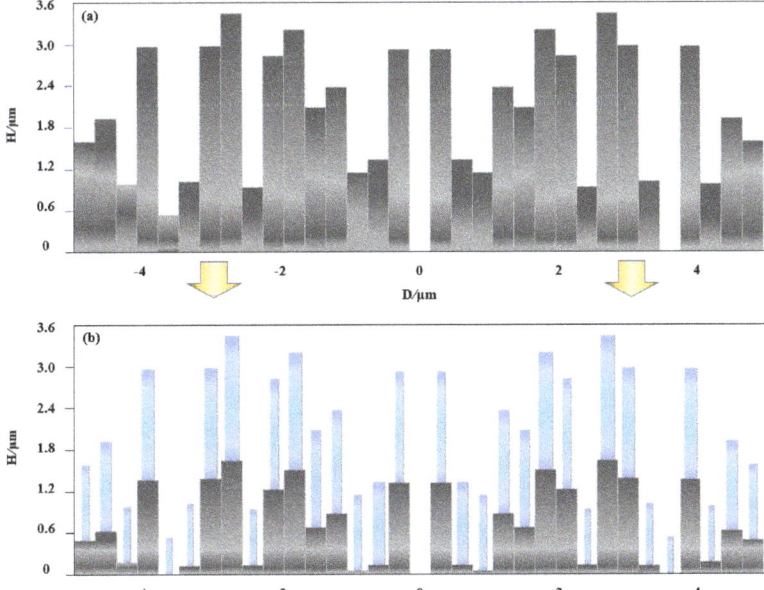

**Figure 4.** Height distribution of (**a**) MDL and (**b**) MMZP.

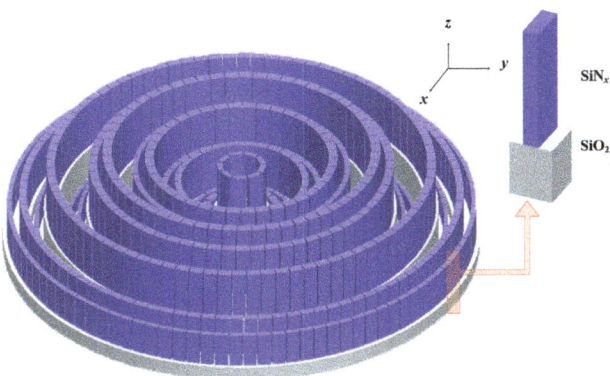

**Figure 5.** The three-dimensional drawing of designed MMZP.

### 3.3. Broadband Focusing and Polariztion Manipulation of the MMZP

In order to verify whether the hybrid design can keep the broadband focusing characteristic and realize the polarization manipulation at the designed wavelength simultaneously, thirty-three kinds of structures are used as meta-atoms to simulate the behavior of the MMZP incident by the circularly polarized light. PML are applied at $x$ or $y$ directions and the propagation direction $z$. The simulated results can be seen in Figure 6.

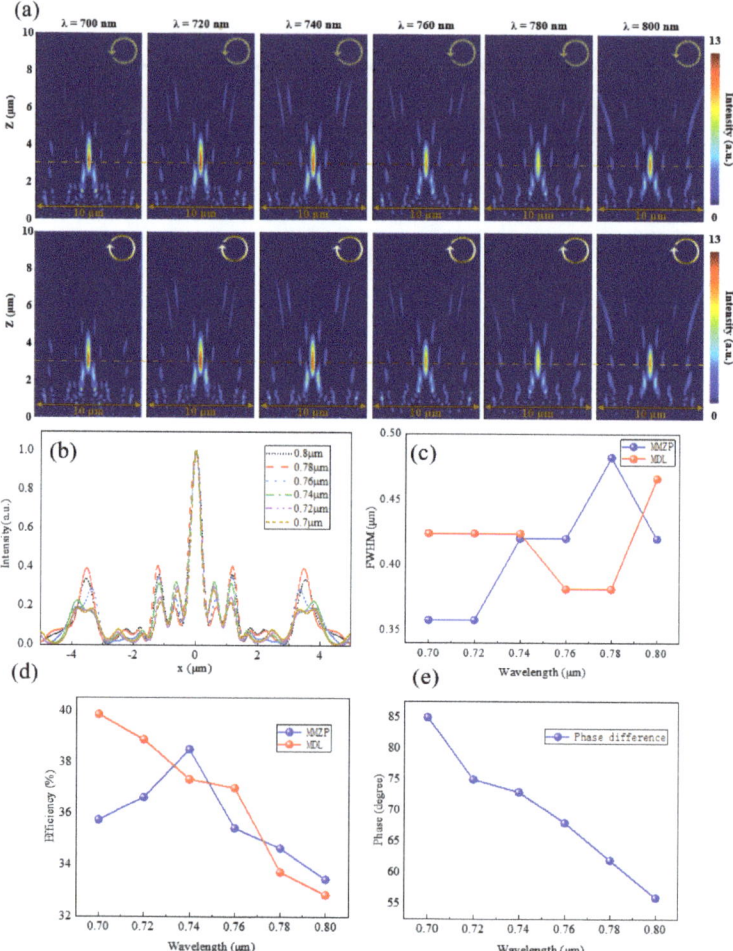

**Figure 6.** Focusing characteristic and polarization manipulation of the MMZP. (**a**) Numerical intensity profiles along axial planes at various incident wavelengths for LCP (top row) and RCP (bottom row) incident light. (**b**) Normalized intensity distribution of the MMZP along transversal axis ($x$ axis). (**c**) FWHM and (**d**) Focusing efficiency at corresponding focal plane as a function of wavelength. (**e**) Simulated phase difference of various incident wavelengths for LCP light.

Figure 6a shows the intensity distribution of the cross section plane. The focal length maintains almost the same while incident light varies, verifying the realization of a near-infrared achromatic broadband focusing feature. Due to the structural characteristics of rotational symmetry, LCP and RCP light have almost the same light field distribution. Figure 6b shows the intensity distributions at the $x$–$z$ plane. The FWHM and the focusing efficiency is shown in Figure 6c,d. All FWHM are nearly half of the corresponding incident wavelength. The focusing efficiency decreased by about 3% compared to MDL, which is due to the reduced duty cycle of the structure resulting in the decrease of energy utilization. Figure 6e shows the results of polarization manipulation. The incident LCP light is set as the superposition of a XLP and YLP light, which has a phase difference of −90°. For the design wavelength of 0.7 µm, the phase difference is close to 90°, which meets our expectation. It means the incident light can be converted from left circularly polarized light to right circularly polarized light after passing through the MMZP. It can be seen that the hybrid design can realize the polarization manipulation while keeping the broadband achromatic in the near-IR.

### 3.4. Polarization-Insensitive Feature of the MMZP

The MMZP exhibits insensitivity to the polarization of incident light. Figure 7a,b show the normalized intensity distributions at the $x$–$z$ plane under the incidence of XLP and YLP light, respectively. Figure 7c,d show the normalized intensity distributions at the $x$–$z$ plane under the incidence of an arbitrary linearly polarized light (45°) and the elliptically polarized incident light. The MMZP maintains the broadband achromatic property when the incident light is in different polarization states.

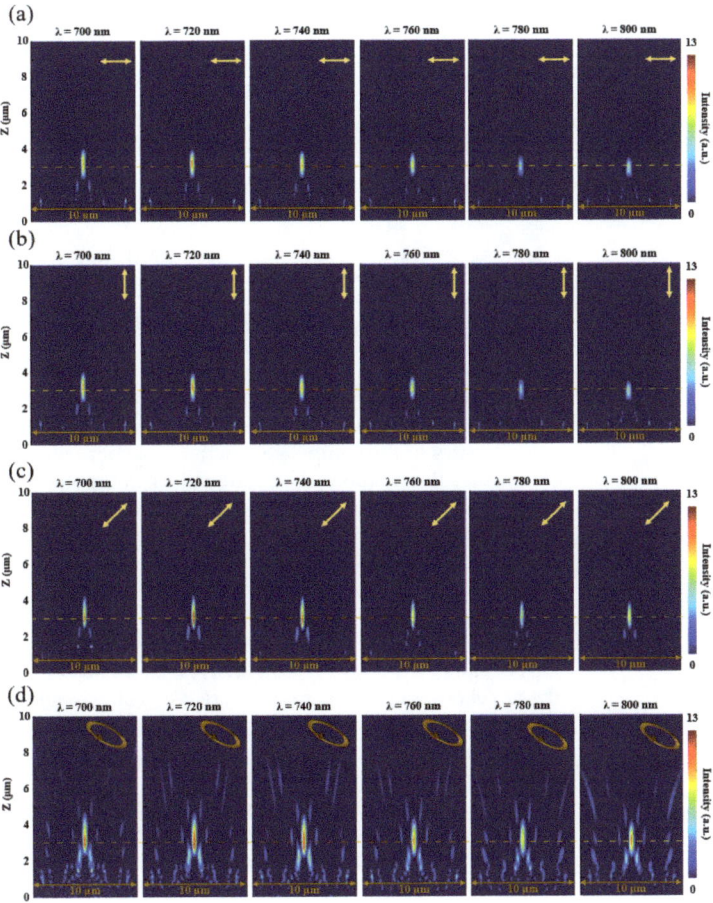

**Figure 7.** Numerical intensity profiles along axial planes at various incident wavelengths for (**a**) XLP, (**b**) YLP, (**c**) 45° linearly polarized (45°-LP) and (**d**) elliptically polarized (EP) incident light.

Figure 8 shows the FWHM of 0.74 μm wavelength with the different polarization states. After passing through the MMZP, the elliptically polarized incident light has the strongest light intensity, followed by the linearly polarized light (45°). The XLP and YLP light has the weakest light intensity. The results indicate the polarization-insensitive feature of MMZP.

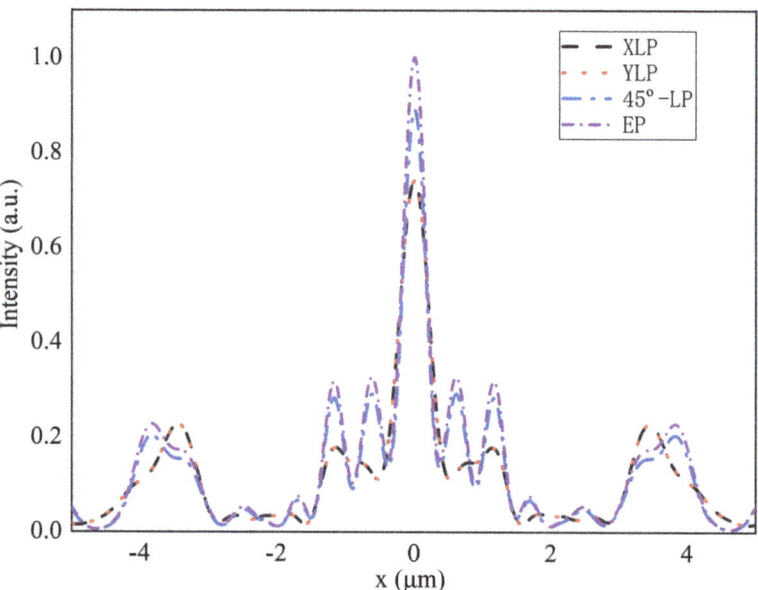

**Figure 8.** Numerical intensity profiles along axial planes at various incident wavelengths for Intensity distributions of the corresponding focal plane.

It can be seen that the LCP light can be converted into RCP light while achieving efficient achromatic broadband focusing under normal incidence. In fact, by adjusting the geometry of the composite structure, the light wave of any polarization state can be obtained by this method [24]. Based on this, the application of MMZP in polarization imaging is the next step to be carried out. Besides, the refractive index of the coating changes with the temperature. Therefore, temperature certainly has an effect on focusing and even polarization modulation. Discussion of the relevance of the temperature dependence of the coating is expected to be carried out in the further research.

## 4. Conclusions

In summary, unlike typical MZPs, which use metasurface instead of diffraction rings, the novel nested composite structured MMZP was successfully modeled by integrating metasurface on the surface of the MDL rings. Based on the global optimization mathematical iterative method, the height distribution of the MMZP is optimized to realize the highly efficient achromatic broadband focusing. The focal length maintains almost the same while incident wavelength varying from 0.7 to 0.8 μm, indicating the realization of a broadband achromatic converging property in near-IR region. The focusing efficiency reaches above 35%, which is more than twice as much as reported in the published results. Furthermore, the polarization state of incident light is accurately regulated by scanning and iterating the dimension parameters of the composite structure. These results indicate that the MMZP has promising practical application prospects in multifunctional focusing devices and polarization imaging.

**Author Contributions:** Conceptualization, S.G. and W.L.; methodology, S.G.; software, S.G., X.S. and X.P.; formal analysis, S.G.; investigation, S.G., P.Y., S.Z. and Y.Z.; data curation, S.G.; writing—original draft preparation, S.G.; writing—review and editing, J.Z.; visualization, S.G.; funding acquisition, S.G. and Y.X. All authors read and agreed to the published version of the manuscript.

**Funding:** This research was funded by the China National Key R&D Program (Grant No. 2018YFE 0199200), the Natural Science Basic Research Program of Shaanxi (Grant No. 2020JQ-805), the Xi'an Key Laboratory of Intelligent Detection and Perception (Grant No. 201805061ZD12CG45), the Key Industry Innovation Chain Project of Shaanxi Provincial Science and the Technology Department (Grant No. 2018ZDCXL-GY-08-02-01), and the School of Optoelectronic Engineering Dean's fund (Grant No. 2019GDYJY02).

**Data Availability Statement:** Data are contained within the article. The data presented in this study are available in Section 2 (Methods) and Section 3 (Results and Discussion).

**Conflicts of Interest:** The authors declare no conflict of interest.

## References

1. Rayleigh, L. *On the Scattering of Light by Small Particles*, 1st ed.; Philosophical Magazine: London, UK, 1871; pp. 447–454.
2. Wood, R. Phase-reversal zone-plates, and diffraction-telescopes. *Lond. Edinb. Dublin Philos. Mag. J. Sci.* **1898**, *45*, 511–522. [CrossRef]
3. Veldkamp, W.; Swanson, G.; Shaver, D. High efficiency binary lenses. *Opt. Commun.* **1985**, *53*, 353–358. [CrossRef]
4. Wood, A.P. Hybrid refractive-diffractive lens for manufacture by diamond turning. In *Commercial Applications of Precision Manufacturing at the Sub-Micron Level, London, UK, 1 April 1992*; SPIE: Bellingham, WA, USA, 1992.
5. Morgan, B.; Waits, C.M.; Krizmanic, J.; Ghodssi, R. Development of a deep silicon phase Fresnel lens using gray-scale lithography and deep reactive ion etching. *J. Microelectromech. Syst.* **2004**, *13*, 113–120. [CrossRef]
6. Li, W.; He, P.; Yuan, W.; Yu, Y. Efficiency-enhanced and sidelobe-suppressed super-oscillatory lenses for sub-diffraction-limit fluorescence imaging with ultralong working distance. *Nanoscale* **2020**, *12*, 7063–7071. [CrossRef]
7. Li, W.; Yu, Y.; Yuan, W. Flexible focusing patterns realization of centimeter-scale planar super-oscillatory lenses in parallel fabrication. *Nanoscale* **2019**, *11*, 311–320. [CrossRef]
8. Yu, Y.; Li, W.; Li, H.; Li, M.; Yuan, W. Investigation of influencing factors on practical sub-diffraction-limit focusing of planar super-oscillation lenses. *Nanomaterials* **2018**, *8*, 185. [CrossRef] [PubMed]
9. Meem, M.; Majumder, A.; Banerji, S.; Garcia, J.C.; Kigner, O.B.; Hon, P.W.C.; Sensale-Rodriguez, B.; Menon, R. Imaging from the visible to the longwave infrared wavelengths via an inverse-designed flat lens. *Opt. Express* **2021**, *29*, 20715–20723. [CrossRef]
10. Meem, M.; Banerji, S.; Majumder, A.; Vasquez, F.G.; Sensale-Rodriguez, B.; Menon, R. Broadband lightweight flat lenses for long-wave infrared imaging. *Proc. Natl. Acad. Sci. USA* **2019**, *116*, 21375–21378. [CrossRef]
11. Banerji, S.; Meem, M.; Majumder, A.; Vasquez, F. G.; Sensale-Rodriguez, B.; Menon, R. Ultra-thin near infrared camera enabled by a flat multi-level diffractive lens: Erratum. *Opt. Lett.* **2019**, *45*, 3183–3183. [CrossRef] [PubMed]
12. Meem, M.; Banerji, S.; Pies, C.; Oberbiermann, T.; Majumder, A.; Sensale-Rodriguez, B.; Menon, R. Large-area, high-numerical-aperture multi-level diffractive lens via inverse design. *Optica* **2020**, *7*, 252–253. [CrossRef]
13. Banerji, S.; Meem, M.; Majumder, A.; Sensale-Rodriguez, B.; Menon, R. Extreme-depth-of-focus imaging with a flat lens. *Optica* **2020**, *7*, 214–217. [CrossRef]
14. Li, S.; Li, X.; Zhang, L.; Wang, G.; Zhang, L.; Liu, M.; Zeng, C.; Wang, L.; Sun, Q.; Zhao, W. Efficient optical angular momentum manipulation for compact multiplexing and demultiplexing using a dielectric metasurface. *Adv. Opt. Mater.* **2020**, *8*, 1901666. [CrossRef]
15. Zhu, Y.; Yuan, W.; Sun, H.; Yu, Y. Broadband ultra-deep sub-diffraction-limit optical focusing by metallic graded-index (MGRIN) lenses. *Nanomaterials* **2017**, *7*, 221. [CrossRef]
16. Zhu, Y.; Zhou, S.; Wang, Z.; Yu, Y.; Yuan, W.; Liu, W. Investigation on super-resolution focusing performance of a TE-polarized nanoslit-based two-dimensional lens. *Nanomaterials* **2020**, *10*, 3. [CrossRef]
17. Zhu, Y.; Chen, X.; Yuan, W.; Chu, Z.; Wong, K. Y.; Lei, D.; Yu, Y. A waveguide metasurface based quasi-far-field transverse-electric superlens. *Opto-Electron. Adv.* **2021**, *4*, 210013. [CrossRef]
18. Feng, F.; Si, G.; Min, C.; Yuan, X.; Somekh, M. On-chip plasmonic spin-Hall nanograting for simultaneously detecting phase and polarization singularities. *Light Sci. Appl.* **2020**, *9*, 95. [CrossRef] [PubMed]
19. Fu, Y.; Min, C.; Yu, J.; Xie, Z.; Si, G.; Wang, X.; Zhang, Y.; Lei, T.; Lin, J.; Wang, D.; et al. Measuring phase and polarization singularities of light using spin-multiplexing metasurfaces. *Nanoscale* **2019**, *11*, 18303–18310. [CrossRef] [PubMed]
20. Stafeev, S.; Kotlyar, V.; Nalimov, G.; Kotlyar, V.; O'Faolain, L. Subwavelength gratings for polarization conversion and focusing of laser light. *Photonics Nanostruct.-Fundam. Appl.* **2017**, *27*, 32–41. [CrossRef]
21. Liu, T.; Feng, R.; Yi, J.; Burokur, S. N.; Mao, C.; Zhang, H.; Werner, D. H. All-dielectric transformation medium mimicking a broadband converging lens. *Opt. Express* **2018**, *26*, 20331–20341. [CrossRef] [PubMed]

22. Nalimov, A.G.; Kotlyar, V.V. Sharp focus of a circularly polarized optical vortex at the output of a metalens illuminated by linearly polarized light. *Comput. Opt.* **2019**, *43*, 528–534. [CrossRef]
23. Khonina, S.N.; Degtyarev, S.A.; Ustinov, A.V.; Porfirev, A.P. Metalenses for the generation of vector Lissajous beams with a complex Poynting vector density. *Opt. Express* **2021**, *29*, 18634–18645. [CrossRef]
24. Yoon, G.; Jang, J.; Mun, J.; Nam, K.T.; Rho, J. Metasurface zone plate for light manipulation in vectorial regime. *Commun. Phys.* **2019**, *2*, 156. [CrossRef]
25. Li, X.; Tang, J.; Baine, J. Polarization-Independent Metasurface Lens Based on Binary Phase Fresnel Zone Plate. *Nanomaterials* **2020**, *10*, 1467. [CrossRef] [PubMed]
26. Eisenbach, O.; Avayu, O.; Ditcovski, R.; Ellenbogen, T. Metasurfaces based dual wavelength diffractive lenses. *Opt. Express* **2015**, *23*, 3928–3936. [CrossRef]
27. Williams, C.; Montelongo, Y.; Wilkinson, T.D. Plasmonic Metalens for Narrowband Dual-Focus Imaging. *Adv. Opt. Mater.* **2017**, *5*, 1700811. [CrossRef]
28. Avayu, O.; Almeida, E.; Prior, Y.; Ellenbogen, T. Composite functional metasurfaces for multispectral achromatic optics. *Nat. Commun.* **2017**, *8*, 14992. [CrossRef]
29. Cai, H.; Czaplewski, D.; Ogando, K.; Martinson, A.; Gosztola, D.; Stan, L.; López, D. Ultrathin transmissive metasurfaces for multi-wavelength optics in the visible. *Appl. Phys. Lett.* **2019**, *114*, 071106. [CrossRef]
30. Minerbi, E.; Keren-Zur, S.; Ellenbogen, T. Nonlinear metasurface fresnel zone plates for terahertz generation and manipulation. *Nano Lett.* **2019**, *19*, 6072–6077. [CrossRef] [PubMed]
31. Hu, Y.; Liu, X.; Jin, M.; Tang, Y.; Zhang, X.; Li, K.F.; Zhao, Y.; Li, G.; Zhou, J. Dielectric metasurface zone plate for the generation of focusing vortex beams. *PhotoniX* **2021**, *2*, 10. [CrossRef]
32. Lou, K.; Qian, S. X.; Ren, Z. C.; Tu, C.; Li, Y.; Wang, H. T. Femtosecond laser processing by using patterned vector optical fields. *Sci. Rep.* **2017**, *3*, 2281. [CrossRef]
33. Danilov, P.A.; Saraeva, I.N.; Kudryashov, S.I.; Porfirev, A.P.; Kuchmizhak, A.A.; Zhizhchenko, A.Y.; Rudenko, A.A.; Umanskaya, S.F.; Zayarny, D.A.; Ionin, A.A.; et al. Polarization-selective Excitation of Dye Luminescence on a Gold Film by Structured Ultrashort Laser Pulses. *JETP Lett.* **2017**, *107*, 15–18. [CrossRef]
34. Porfirev, A.; Khonina, S.; Meshalkin, A.; Ivliev, N.; Achimova, E.; Abashkin, V.; Prisacar A.; Podlipnov, V. Two-step maskless fabrication of compound fork-shaped gratings in nanomultilayer structures based on chalcogenide glasses. *Opt. Lett.* **2017**, *46*, 3037–3040. [CrossRef]
35. Wang, P.; Menon, R. Computational spectrometer based on a broadband diffractive optic. *Opt. Express* **2014**, *22*, 14575–14587. [CrossRef] [PubMed]
36. Wang, P.; Mohammad, N.; Menon, R. Chromatic-aberration-corrected diffractive lenses for ultra-broadband focusing. *Sci. Rep.* **2016**, *6*, 21545. [CrossRef]
37. Kim, G.; Domínguez-Caballero, J.A.; Menon, R. Design and analysis of multi-wavelength diffractive optics. *Opt. Express* **2012**, *20*, 2814–2823. [CrossRef] [PubMed]
38. Li, S.; Li, X.; Wang, G.; Liu, S.; Zhang, L.; Zeng, C.; Wang, L.; Sun, Q.; Zhao, W.; Zhang, W. Multidimensional manipulation of photonic spin Hall effect with a single-layer dielectric metasurface. *Adv. Opt. Mater.* **2019**, *7*, 1801365. [CrossRef]
39. Ge, S.; Liu, W.; Zhang, J.; Huang, Y.; Xi, Y.; Yang, P.; Sun, X.; Li, S.; Lin, D.; Zhou, S. Novel Bilayer Micropyramid Structure Photonic Nanojet for Enhancing a Focused Optical Field. *Nanomaterials* **2021**, *11*, 2034. [CrossRef] [PubMed]
40. Ge, S.; Liu, W.; Zhou, S.; Li, S.; Sun, X.; Huang, Y.; Yang, P.; Zhang, J.; Lin, D. Design and preparation of a micropyramid structured thin film for broadband infrared antireflection. *Coatings* **2018**, *8*, 192. [CrossRef]

Article

# Facilely Flexible Imprinted Hemispherical Cavity Array for Effective Plasmonic Coupling as SERS Substrate

Jihua Xu [1], Jinmeng Li [1], Guangxu Guo [1], Xiaofei Zhao [1], Zhen Li [1], Shicai Xu [2], Chonghui Li [2], Baoyuan Man [1], Jing Yu [1,*] and Chao Zhang [1,*]

[1] Collaborative Innovation Center of Light Manipulations and Applications, Institute of Materials and Clean Energy, School of Physics and Electronics, Shandong Normal University, Jinan 250358, China; xujihua23@163.com (J.X.); jinmeng345@163.com (J.L.); czsdnu@hotmail.com (G.G.); zxfsdnu@126.com (X.Z.); lizhen19910528@163.com (Z.L.); manyuanman_sdnu@163.com (B.M.)
[2] Shandong Key Laboratory of Biophysics, Institute of Biophysics, Dezhou University, Dezhou 253023, China; shicaixu@dzu.edu.cn (S.X.); chonghuili163@163.com (C.L.)
* Correspondence: yujing1608@126.com (J.Y.); czsdnu@126.com (C.Z.)

**Abstract:** The focusing field effect excited by the cavity mode has a positive coupling effect with the metal localized surface plasmon resonance (LSPR) effect, which can stimulate a stronger local electromagnetic field. Therefore, we combined the self-organizing process for component and array manufacturing with imprinting technology to construct a cheap and reproducible flexible polyvinyl alcohol (PVA) nanocavity array decorating with the silver nanoparticles (Ag NPs). The distribution of the local electromagnetic field was simulated theoretically, and the surface-enhanced Raman scattering (SERS) performance of the substrate was evaluated experimentally. The substrate shows excellent mechanical stability in bending experiments. It was proved theoretically and experimentally that the substrate still provides a stable signal when the excited light is incident from different angles. This flexible substrate can achieve low-cost, highly sensitive, uniform and conducive SERS detection, especially in situ detection, which shows a promising application prospect in food safety and biomedicine.

**Keywords:** surface-enhanced Raman spectroscopy; localized surface plasmon resonance; flexible SERS substrate; cavity

**Citation:** Xu, J.; Li, J.; Guo, G.; Zhao, X.; Li, Z.; Xu, S.; Li, C.; Man, B.; Yu, J.; Zhang, C. Facilely Flexible Imprinted Hemispherical Cavity Array for Effective Plasmonic Coupling as SERS Substrate. *Nanomaterials* **2021**, *11*, 3196. https://doi.org/10.3390/nano11123196

Academic Editor: Onofrio M. Maragò

Received: 10 November 2021
Accepted: 24 November 2021
Published: 25 November 2021

**Publisher's Note:** MDPI stays neutral with regard to jurisdictional claims in published maps and institutional affiliations.

**Copyright:** © 2021 by the authors. Licensee MDPI, Basel, Switzerland. This article is an open access article distributed under the terms and conditions of the Creative Commons Attribution (CC BY) license (https://creativecommons.org/licenses/by/4.0/).

## 1. Introduction

The Raman spectrum has been widely used in qualitative and quantitative analysis as a nondestructive analysis tool, which can provide molecular vibration information [1]. However, the weak signal due to the low Raman cross-section area greatly hinders the development of this technology [2,3]. Surface-enhanced Raman scattering (SERS) has been widely used in food and environmental safety, early disease diagnosis, drug detection, in situ monitoring of biological environment and chemical reactions, which can efficiently enhance the Raman signal [4–8]. As we all know, the amplification of the Raman signal is usually realized by an electromagnetic mechanism that is mainly depends on the localized surface plasmon resonance (LSPR) effect whereby collective oscillation of the metal's conduction band electrons is excited upon light excitation [9]. It is reported that the electromagnetic field intensity is inversely proportional to the distance parameter, which means that the effective SERS signal is generated on the premise that the molecule is located in the confined area around the plasmonic material [10]. Various types of SERS substrate structures ranging from nanoparticle (NP) aggregates [11–15], multidimensional structure [16–20] and composite substrate [21–29] were prepared to explore the better enhancement factor (EF) under various conditions. Among them, the cavity system has been paid more and more attention because of its unique light trapping characteristics [30–38]. The combination of surface and cavity plasmonic modes can provide an effective way for electromagnetic field enhancement. For instance, alumina nanocavity arrays were prepared by drying and decomposing the aqueous solution of Al $(NO_3)_3$ spun on the monolayers of PS spheres [39], which provided the distribution of hotspots in three-dimensional space. Compared with uncomplicated structures (such as nanoparticles and films), the light reflection caused by gradient refractive index and multiple internal scattering can be effectively suppressed in metal nanocavities [40]. Inspired by the equivalence of light propagation between media containing gradients in optical properties and warped geometries of space-time, Peng Mao et al. designed and proved that warped spaces can be used to introduce extremely localized energy [41]. More recently, Zewen Zuo et al. prepared a polydimethylsiloxane-supported Ag nanocone array covered by gold nanoparticles (Au NP), where multiple types of plasmonic coupling occur in this multi-focus field mode, resulting in significantly enhanced electromagnetic fields and large hot spots area. Generally, ultra-high sensitivity is the basic performance of a superior SERS sensor, but signal stability, uniformity and low production cost in the real-work application are also crucial issues to be investigated [42,43].

Most of the substrate templates used to fabricate cavity structures relied too much on expensive and energy-consuming precision machining processes, such as electron beam lithography and focused ion beam pattern lithography, which greatly limits the practical application of high-performance and reliable SERS substrates. Based on this strategy, we combined the self-assembly technology with the imprinting process to prepare a periodic hemispherical cavity array structure, which can be used as a highly sensitive sensor for transmitting stable, uniform and repeatable signals. Here, we chose PVA as the template layer and the PS microsphere monolayer self-assembled in periodical arrangements as the object to be imprinted to obtain a flexible imprinted hemispherical cavity array. Then, the silver nanoparticles are sputtered in the hemispherical cavity to synergize the metal surface plasmon resonance (LSPR) with the cavity mode, which further enhances the Raman signal. With the aim of calculating the enhancement effect of this substrate, we detected different concentrations of rhodamine 6G (R6G), Crystal Viole (CV) and Malachite green (MG); their Raman signals were enhanced to varying degrees, accompanied by good uniformity. The geometry of metal nanostructures can affect the plasmon resonance effect to a great extent; thus, we compared the different roles of silver film and silver nanoparticles in the cavity. The effects of Ag film and Ag nanoparticles on the strength of hot spots were compared by finite-different time-domain (FDTD) simulation. When silver nanoparticles are orderly attached to the inner wall of the hemispherical cavity, more hot spots are generated between the adjacent nanoparticles. The three-dimensional distribution of hot spots in Ag NP-cavity simulated by COMSOL confirmed this point. In addition, the stability of the substrate was also tested. As the results show, the flexible substrate can be used for quantitative and qualitative analysis and has a wonderful application prospect for reliable in situ detection.

## 2. Materials and Methods

### 2.1. Chemicals and Materials

Acetone ($CH_3COCH_3$), Ethanol ($C_2H_6O$), Sodium dodecyl sulfate (SDS, $C_{12}H_{25}NaO_4S$) and Toluene ($C_7H_8$) were purchased from Sinopharm Chemical Reagent Co., Ltd. (Shanghai, China). Monodisperse polystyrene (PS) sphere suspensions (5 wt% in water) with diameters of 500 nm were purchased from Maxinne dareen (China). Single-sided polished silicon wafers were purchased from Lijing Technology Co., Ltd. (Shenzhen, China). PVA 1788 (87.0–89.0% alcoholysis degree) and malachite green (MG, MW:463) were purchased from Shanghai Aladdin Biochemical Technology Co., Ltd. (Shanghai, China). Rhodamine 6G (R6G, MW:479) was purchased from Meilun Biotechnology Co., Ltd. (Dalian, China). Crystal violet (CV, MW:407) was purchased from Yuanye Biotechnology Co., Ltd. (Shanghai, China).

*2.2. Preparation of Tightly Packed Polystyrene Sphere Monolayer*

PS microsphere arrays were prepared by gas-water interface self-assembly technology: The cleaned slides were firstly treated by oxygen plasma for 120 s to obtain a highly hydrophilic surface. Subsequently, after mixing ethanol and PS microspheres at the ratio of 2:1 and ultrasonic treatment of 15 min, a small amount of the mixture was dripped on the slide, covering the entire upper surface of the slide. Then, it was left at room temperature for an hour until a large area of the continuous molecular layer of PS microspheres was formed on the surface of the slide. Sodium dodecyl sulfate with a concentration of 2 wt% was prepared as a surfactant drop on the surface of deionized water (DI) to change the surface tension of the DI and drive the nanospheres into a hexagonal and tightly packed monolayer [44]. The PS microsphere monolayer on the glass slide was detached from the deionized water surface in the beaker, and then the molecular film was obtained from the silicon wafer, which was also treated by oxygen plasma; thus, the self-assembly monolayer of the PS microsphere was transferred to the silicon wafer.

*2.3. Preparation of Ag NPs/PVA Nano-Bowl Cavity Array Substrate*

In order to make the PS microspheres firmly adhere to the silicon wafer surface, we heated the PS spherical monolayer on the silicon wafer surface to 20 min at 90 °C PVA aqueous solutions with different concentrations were prepared by stirring the mixture of polyvinyl alcohol powder and deionized water at 80 °C for 6 h, and then spin-coated on the monolayer of PS spheres by a homogenizing machine and dried at 60 °C for 40 min to obtain sandwich structure on silicon wafers. With the transparency of PVA in mind, we stuck a tiny label film that is chemically stable and has no effect on the substrate to the corner of the sample to distinguish between the positive and negative sides of the substrate. Before completely soaking and etching the PS microspheres in toluene solution, the silicon wafers were peeled off to obtain the structure of a single-layer PS microsphere array covered with PVA. After soaking in toluene solution for 40 h, the PVA flexible substrate with sunken bowl-shaped cavity array (PC) was fabricated. Silver nanoparticles were Sputtered on the PVA bowl cavity structure by the Denton Vacuum DESKTOP PRO turbo pumped confocal sputter deposition system, and the size of Ag NPs was changed by setting different times. The preparing technology was completed when the sample was coated with an ultra-thin gold film (3 nm) to slow down the oxidation reaction, the Ag NPs/PVA bowl-shaped cavity array composition was obtained.

*2.4. Characterization*

The morphologies and compositions of the prepared samples were characterized by scanning electron microscopy (SEM, Zeiss Gemini Ultra-55, Jena, Germany) equipped with energy-dispersive X-ray spectroscopy. SERS spectra were measured using the Horiba HR Evolution 800 Raman spectrometer (Kyoto, Japan). Raman measurement conditions: laser power = 0.48 mW; diffraction grid = 600 g/nm; objective lens = 50×; laser wavelength = 532 nm. All the detected molecules were prepared by dissolving them into ethanol solution.

*2.5. Theoretical Simulation*

The electromagnetic field distributions of Ag NPs/PVA nano-bowl cavity (Ag NP-on-PC) and Ag film-on-PC structures were first simulated by FDTD, and the $x$-$z$ cross-sectional views were compared. In order to visualize the distribution of hot spots in the 3D nanocavity, we then used commercial COMSOL Multiphysics software to simulate the distribution of silver nanoparticles in the PVA cavity. To control the uniqueness variables, in all models, the diameters of silver nanoparticles and hemispherical cavities were set to 35 nm and 500 nm, and the nano-gap between the particles was set to 3 nm. The polarization direction was set along the $x$-axis. In all the directions, the absorption boundary condition was the perfect matching layer (PML). The incident light with the wavelength of 532 nm was used and incident along the $z$-direction. The corresponding refractive index of PVA was set as 1.4835 and the dielectric functions of Ag were taken from Palik. The refractive index of the surrounding medium was 1.0 for air.

## 3. Results

### 3.1. Morphologies of PVA Nanocavity Arrays and the Effect of the Concentration of PVA

The preparation process of the 3D flexible Ag NPs/PVA bowl-shaped cavity array substrate is illustrated schematically in Figure 1. More information about the synthesis process and SERS spectrum measurement is shown in the Materials and Methods. During the preparation of the PVA flexible nano-bowl array template, we found that different morphologies could be obtained by covering different concentrations of polyvinyl alcohol on the monolayer PS microsphere array. The SEM images of these structures are shown for comparison. The morphology dependence of the concentration of PVA solution can be understood from the fluidity before it is fixed to form hemispherical cavities. When the concentration is 10 wt% (Figure 2a), the extremely fluid PVA permeates every gap between the microspheres, leaving a double-layer solidified reticular structure of PVA after the microspheres are completely corroded. When the concentration of PVA increases to 15 wt%, the PVA solution will not flow to the bottom of some PS microspheres, and the monolayer nano-bowl structure begins to appear. As the concentration increases, the fluidity of the PVA solution weakens, the area of double-layer reticular decreases, and the substrate gradually tends to be uniform. When the concentration of PVA reaches 25 wt%, we obtain a large area of uniform and tightly arranged bowl-shaped cavity array structure as shown in Figure 2d. The efficient and intuitive views illustrate the effect of the concentration of PVA solution on the morphology.

In order to endow SERS activity, silver nanoparticles were sputtered on the surface of the obtained PVA flexible nano-bowl array template. Subsequently, we separately detected the R6G molecule with a concentration of $10^{-6}$ M on these different structures, and the results are shown in Figure 3. Apparently, the double-decker covered with a reticular composition exhibits relatively poor SERS activity. As can be seen in the illustration in Figure 2a–c, the diameter of the upper meshwork is smaller than the underlying hemispherical cavity so that the silver electron beams are blocked and composed of nanoparticles on the upper meshwork. The number of Ag nanoparticles attached to the inner wall of the hemispherical cavity is much lower; thereupon, the Ag NP-cavity system cannot be formed, which weakens the coupling effect between metal plasma and intracavitary-focusing field, the Raman performance was greatly reduced. In contrast, the remarkably strongest Raman signal was detected on the Ag-PC-25 substrate, indicating that the pure hemispherical cavity structure is more conducive to coupling.

### 3.2. SERS Performance and the Dependence of the Ag NPs/PVA Nanocavity Array

Figure 4a presents the Raman spectra of the $10^{-6}$ M R6G alcoholic solution detected on fabricated Ag NPs/PVA nanocavity arrays, which were obtained by sputtering silver with different duration, where the change in the Raman signal intensity with increasing the deposition time from 20 s to 80 s is exhibited. Evidently, the optimal enhancement effect on the Raman signal can be achieved when the sputtering time is 70 s. The intensities of the Raman peaks at 613, 1365 and 1651 cm$^{-1}$ plotted as functions of the sputtering time are shown in Figure 4b, from which we can observe that the Raman intensities increase with Ag sputtering time increasing until 70 s, and then decrease when the time continues to increase. The SERS activity dependence of the sputtering time can be understood from the coupling strength of the cavity mode with the metal particle plasma. In the process of deposition, 3D-coated silver nanoparticles instead of continuous films are produced in the early stage due to the limited wettability of the substrate [45]. When the sputtering time is not long enough, the small size of the nanoparticles not only leads to the weak LSPR effect but also weakens the focusing field effect because the cavity film is not formed. When the sputtering time is appropriate, a sufficient amount of LSPR is generated between the silver nanoparticles, and all the particles attached to the bowl-shaped inner wall form a focusing field. There is a strong coupling between the two modes, and with the enhancement of the coupling, a stronger hot spot is produced. However, when the sputtering continues after the "best time", the gap decreases gradually, the hot spot decreases, and the coupling effect weakens. For a homogeneous material, it was proved that the Raman properties of nanoparticles on planar substrates are much worse than those on warped surfaces because of their uniform refractive index and electromagnetic field, although other variables are consistent [46,47]. To that end, we collected the SERS spectra of R6G over PVA cavity, Ag NP-on-PC, Ag film-on-PC and continuous Ag film substrates in addition. As illustrated in Figure 4c, all the characteristic Raman peaks of R6G can be detected in these four cases, in which the enhancement effect of the PVA cavity was measured with R6G at a concentration of $10^{-2}$ M, and R6G at the concentration of $10^{-6}$ M is tested for the other substrates. Obviously, it can be seen the enhancement effect of Ag NP-on-PC structure is much stronger than that of that on FP, which is attributed to the cavity mode. It is expected that the higher signal strength exhibited on Ag NP-on-PC, whether compared to Ag film-on-PC or nudity PVA cavity, provides further evidence of the high-density hot spots generated because of the synergy.

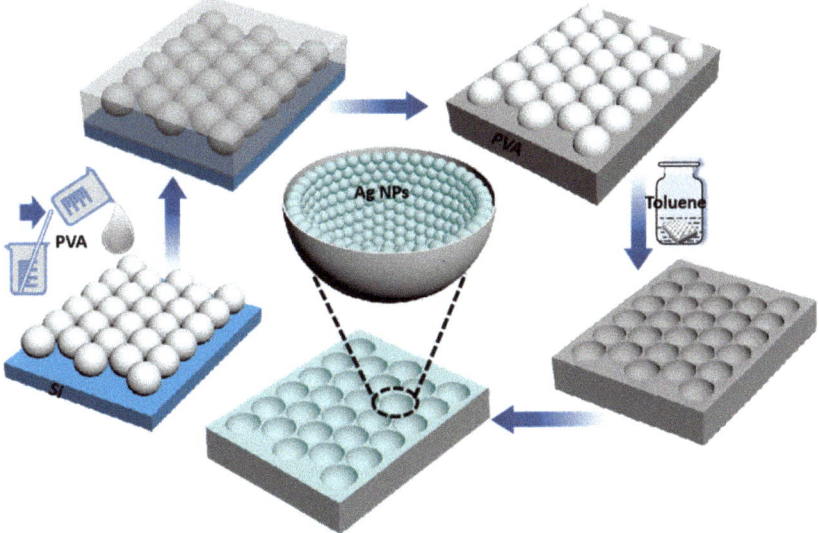

**Figure 1.** The fabrication process of hierarchical Ag NPs/PVA bowl-shaped cavity array nanostructure.

We implement 3D simulations of hot spots among nanoparticles in the cavity by using COMSOL. Before establishing the model, we calculated a statistical analysis of the diameter distribution of Ag NPs based on SEM images of Ag NP-on-FP obtained in the same experimental environment (Figure 5a), acquiring a logarithmic normal distribution with a mean value of 35 nm and a standard deviation of 5 nm, as shown as in Figure 5b. In Figure 5c, it can be seen that the hot spots excited by the LSPR effect are 3D-distributed between the adjacent particles. In order to further demonstrate the influence of silver nanoparticles and silver film on the focusing field effect, we simulated the Ag NP-on-PC and Ag film-on-PC structures. The incident light is concentrated and forms a strong focusing field in the hemispherical cavity under 532 nm light excitation. The local electric field distributions at x-z cross-section of these two structures were shown in Figure 5d,e, where it can be observed that the Ag NP-cavity can produce more local electromagnetic fields (hot spots) distributed in the narrow gaps between the silver nanoparticles compared with the Ag film-cavity. In addition, the weak reflection ability caused by its discontinuity produces a weak focus field similar to the latter. On the other hand, the silver film cavity can excite a stronger focusing field mode; even so, its overall enhancement effect is not comparable to that of silver nanoparticles. As we know, the metal nanoparticles themselves can produce a higher local electric field, and there is coupling between the particles, resulting in more hot spots. Because the focusing field has a larger electromagnetic (EM) energy density than the incident plane wave, the plasmon resonance excited by the focusing field makes the local field of the nanoparticles stronger, thus strengthening the coupling between the particles. As a result, a further enhanced electric field is generated in the gap between the adjacent particles, which leads to stronger hot spots in the focus field of the Ag NP-cavity, although the focusing field is relatively weak. It was proved that the synergistic effect of plasmon resonance of metal nanoparticles and cavity mode is particularly effective for SERS.

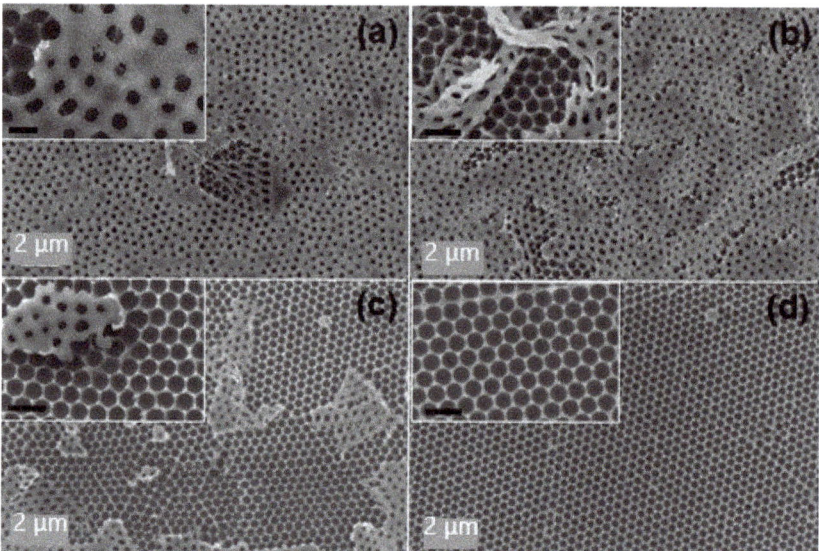

**Figure 2.** Morphology of the fabricated sunken bowl-shaped microcavity arrays, which were obtained by preparing a PVA solution with a concentration of (**a**) 10 wt% (PC-10), (**b**) 15 wt% (PC-15), (**c**) 20 wt% (PC-20) and (**d**) 25 wt% (PC-25), respectively.

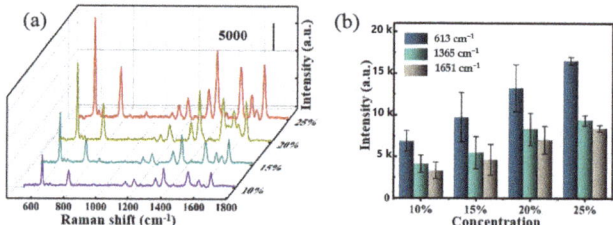

**Figure 3.** (**a**) The Raman spectra of R6G molecules were detected on these four different structures, which were obtained with the PVA concentration of 10 wt% (Ag-PC-10), 15 wt% (Ag-PC-15), 20 wt% (Ag-PC-20) and 25 wt% (Ag-PC-25). The Raman characteristic peak intensities of R6G are shown on (**b**), respectively. It is calculated that RSD = 18.14%, 30.31%, 21.08%, 5.9%, correspondingly.

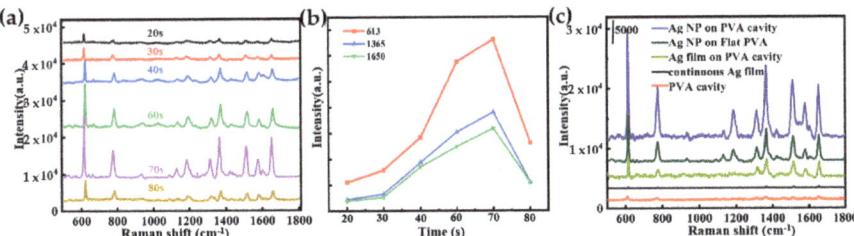

**Figure 4.** (a) SERS spectra of R6G ($10^{-6}$ M) over substrates with different sputtering times were detected, and (b) the intensity of the corresponding characteristic peaks were shown. (c) SERS spectra of R6G ($10^{-2}$ M) over PVA cavity, and the spectra of R6G ($10^{-6}$ M) over Ag NP-on-PC, Ag NP-on Flat PVA(FP), Ag film-on-PC and continuous Ag film substrates.

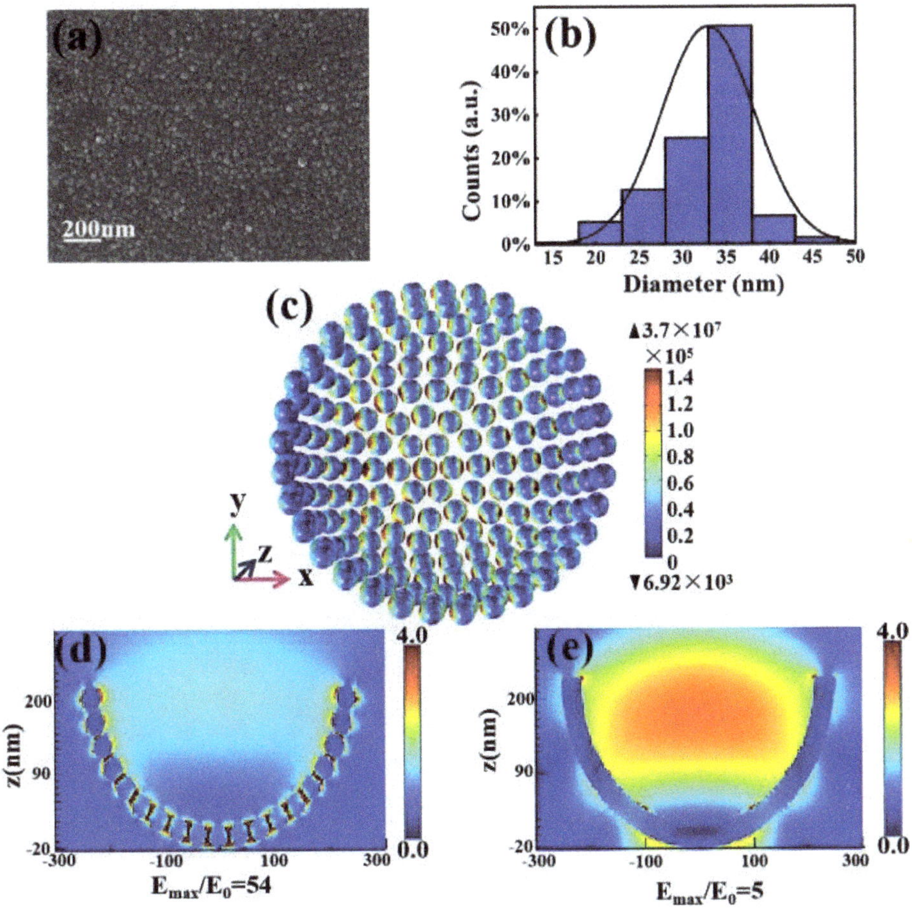

**Figure 5.** (a) SEM images of Ag NPs sputtered on the flat PVA film (Ag NP-on-FP) under the same conditions with Ag NP-on-PC. (b) The distribution of particle diameter is calculated. (c) Simulated electric field distribution of Ag NPs in the inner wall of the cavity and the x-z plane electric field distribution of (d) Ag NP-on-PC and (e) Ag film-on-PC.

## 3.3. Optimized Ag NPs/PVA Nanocavity and SERS Performance

The morphology and structure of the PS microsphere array (Figure 6a), PVA cavity array (Figure 6b) and hemispherical cavity array assembled by Ag nanoparticles (Figure 6c) were characterized by scanning electron microscope (SEM). In Figure 6b, it can be seen that the cavity array was composed of hexagonally arranged hemispherical voids, which replicate the features of the PS microsphere array well. Figure 6c exhibits the morphology of intact substrate, where the hexagonal arrangement and the closely stacked concave array structure were observed in the low power scanning electron microscope images, which was uniformly covered with silver nanoparticles, and the average size was about 30 nm. EDS elemental mappings (Figure 6d) of local samples showed that carbon (purple), oxygen (yellow) and silver (orange) elements were uniformly distributed on the tested samples. The large-scale highly ordered inverse PS hemispherical monolayer template ensures the successful preparation of the hemispherical cavity array and further ensures the high uniformity of SERS spectra at different positions on the whole substrate. This simple method makes the silver hemispherical cavity array assembled by different batches of silver nanoparticles have high structural repeatability, thus ensuring the high repeatability of SERS spectra. In addition, we also measured the UV–vis spectra of Ag NPs/PVA nanocavity and pure PVA film (Figure S1a,b), and it can be seen that composite substrate maintains a higher absorption of light with the band from 200 to 500 nm, which is related to its low reflectivity in this band as shown in Figure S1b.

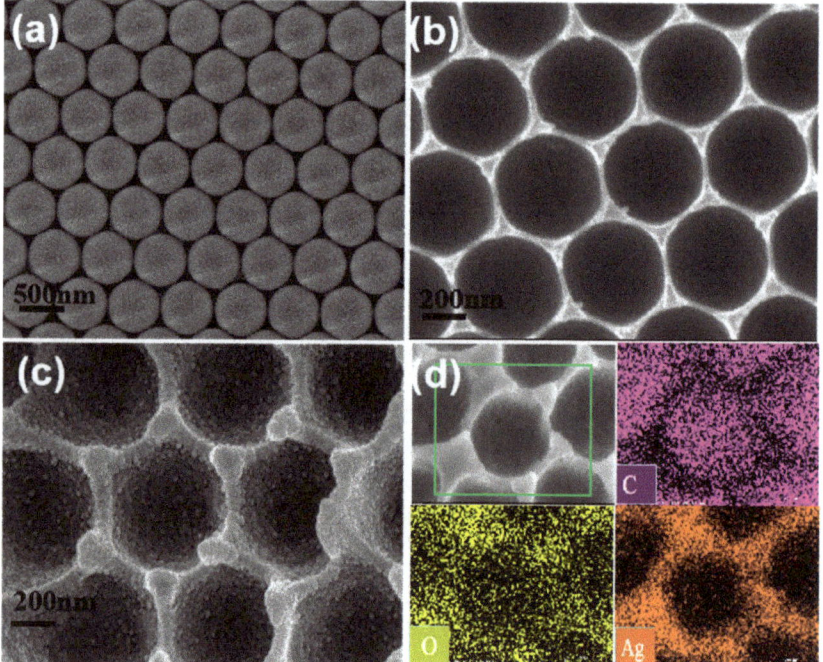

**Figure 6.** Morphology characterization from PS microsphere array (**a**), hemispherical cavity array (**b**), Ag NPs/PVA cavity composite substrate (**c**), respectively. (**d**) EDS elemental maps from C, O and Ag on (**c**).

In order to evaluate the Raman properties of the Ag NPs/PVA cavity composite structure, R6G, MG and CV alcoholic solution were prepared to be detected as probe molecules, and the SERS spectra of them are shown in Figure 7a,d,e, respectively. Apparently, the intensity of the Raman signal lessens gradually with the decrease in the concentration of probe molecules. Figure 7a shows the SERS spectra of R6G alcoholic solution with concentrations ranging from $10^{-5}$ to $10^{-10}$ M. It can be observed that distinguishable peaks are demonstrated even at diluted concentrations of $10^{-10}$ M, and the limit of detection (LOD) of this flexible substrate is obtained. In order to quantize the SERS performances, the enhanced factor (EF) was evaluated using the formula [48]:

$$EF = \frac{I_{SERS}/N_{SERS}}{I_{RS}/N_{RS}}$$

where $I_{SERS}$ and $I_{RS}$ represent the intensity of SERS spectra and normal Raman and $N_{SERS}$ and $N_{RS}$ refer to the average number of molecules within the laser spot excited by SERS and normal Raman, respectively. Here, the value of $N_{RS}/N_{SERS}$ was estimated with the ratio of the respective molecule concentrations. The peak strength of $10^{-2}$ M R6G at 613 cm$^{-1}$ on PVA is 232, and the peak intensity of $10^{-10}$ M R6G on Ag NPs/PVA cavity composite structure is 637 (Figure S2). As a result, the average EF is $2.75 \times 10^8$, which proves that this substrate has high SERS activity. In order to examine the ability of quantitative detection, the experimental values were logarithmically plotted to evaluate the correlation between concentration and relative intensity at 613 and 771 cm$^{-1}$, and then the linear fitting curve (two solid lines) with a high coefficient of determination ($R^2_{613} = 0.991$, $R^2_{771} = 0.983$) was obtained, as shown in Figure 7b. Repeatability and universality are two indispensable indicators to measure the practical application of the SERS sensor.

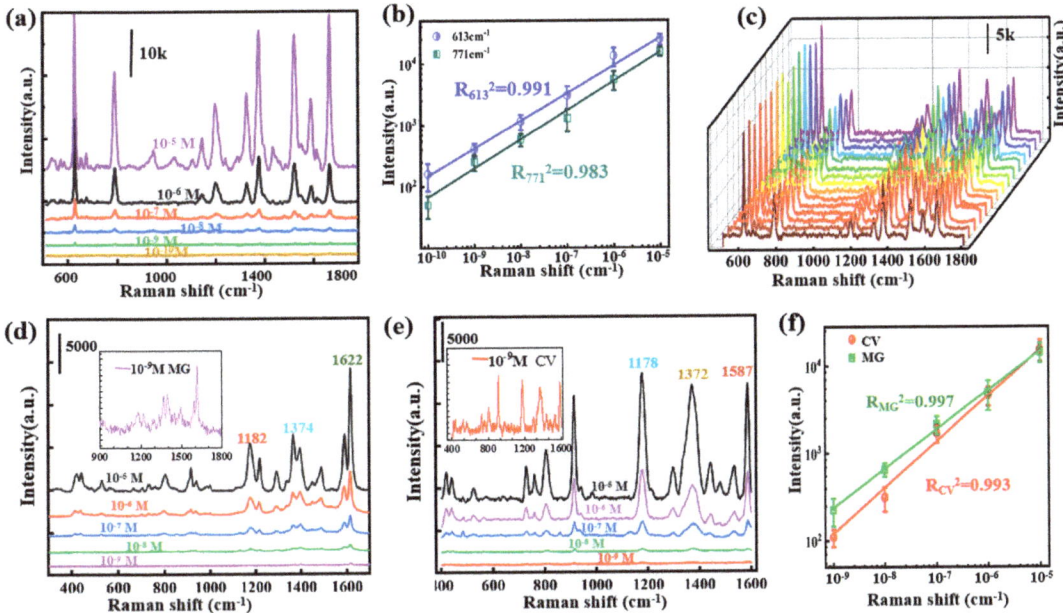

**Figure 7.** (a) SERS-normalized spectra of R6G on Ag NPs/PVA cavity composite structure at varying concentrations from $10^{-5}$ to $10^{-10}$ M. (b) Linear relationships: Raman intensities at 613 and 771 cm$^{-1}$ as a function of the concentrations of R6G molecules on Ag NPs/PVA cavity composite structure. (c) The waterfall map consists of Raman spectra obtained by detecting target molecules with different batches of samples (Ag NPs/PVA cavity composite structure). (d,e) SERS spectra of MG and CV at varying concentrations from $10^{-5}$ to $10^{-9}$ M detected on the Ag NPs/PVA composite substrate. Insets in (d,e) are the enlarged SERS spectrum of $10^{-9}$ M MG and CV, respectively. (f) Plots of the logarithmic integrated intensities of the Raman peaks at 1372 cm$^{-1}$ for CV and 1622 cm$^{-1}$ for MG as functions of the logarithmic concentration, and the solid lines are the linear fittings of the data.

Figure 7c presents the Raman spectra of different batches of R6G with the concentration as $10^{-6}$ M confirming the reliability of the substrate to generate reproducible SERS signals. Then MG and CV molecules with concentrations of $10^{-5}$ to $10^{-9}$ M were also detected; the curves in Figure 7d,e correspond to their Raman spectra in turn. For the spectrum measured at the concentration of $10^{-9}$ M, the main characteristic peaks of MG and CV can still be clearly distinguished, as shown in the illustration. Similarly, the intensity values of the corresponding characteristic peaks of the two substances at each concentration are linearly fitted, and it is calculated that $R^2_{MG} = 0.997$ (green line) and $R^2_{CV} = 0.993$ (red line) as shown in Figure 7f. The results of these linear analyses well prove the quantitative analysis ability of the substrate. In addition, three kinds of probe molecules can be detected quantitatively, which proves that the Ag NPs/PVA composite structure is universal in the practical application of Raman detection.

By taking practical application into account, high sensitivity is only one of the critical aspects that the substrate should provide, and stability signal is also very important. To further examine the application performance of Ag NPs/PVA cavity composite structure, the mechanical stability of the thin film was tested firstly. We bent the substrate repeatedly and caught the real-time response after the substrate was restored to steady state, and the SERS spectra were obtained, as shown in Figure 8a. Figure 8b presents the intensity at 613 cm$^{-1}$ of these Raman spectra. The changing trend of Raman signals with the increase in bending times from these histograms can be more intuitively observed. As we can see, the Raman signal detected has a slight downward trend with the increase in bending times because the micro-deformation of the bowl-shaped cavity changed the gap between the silver nanoparticles attached to the inner wall of the cavity driven by the external force and the hot spot changed accordingly. The "$n$" was defined as the ratio of the reduced peak intensity ($\Delta E$) to its original intensity ($E$), $n_{80} = 7.02\%$, which means that even under the interference of repeated 80 times of high-frequency bending, a slight decrease occurred in peak intensity, which will have a negligible impact on quantitative detection. Furthermore, R6G on Ag NPs/PVA composite structure was detected with laser incident from different angles to investigate the optical stability of the substrate. Generally speaking, strong coupling occurs only when the polarization direction of the excited light is perpendicular to the particle-film interface in the particle-on-film system [49]. As shown in Figure 8c,d, therefore, there is a slight enhancement when the incident angle is 15°. The Raman signal is basically maintained at a stable level with slight fluctuations. In order to further verify this conclusion, we carried out a theoretical simulation, as shown in Figure 8e–j. It can be seen from the simulation results that, as the detected Raman signal shows, the intensity of hot spots changes slightly in a small category. Especially when the incident angle is less than or equal to 60 degrees, the hot spot intensity remains in order of magnitude. Thus far, the mechanical stability and optical stability of the flexible substrate have been verified, which demonstrates that Ag/PVA cavity substrate is a promising candidate for practical curved surface SERS detection.

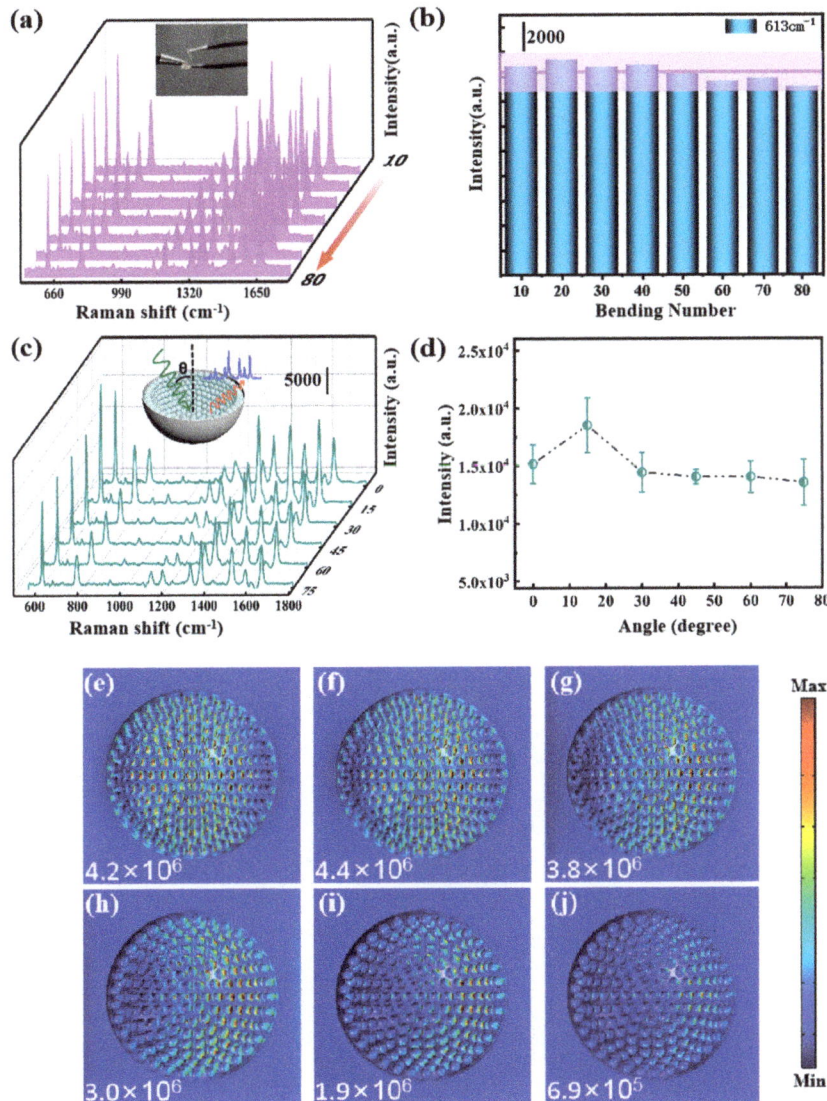

**Figure 8.** (a) Real-time SERS spectra of R6G on Ag NPs/PVA cavity during the bending experiment, and the inset shows the implementation process of bending experiment. (b) The corresponding histogram of intensity distribution of the peak at 613 cm$^{-1}$ (the average intensity is marked with a purple line, and the purple zone represents the ±10% fluctuation). (c) SERS spectra of R6G on Ag NPs/PVA cavity with laser incidence from different angles (0°, 15°, 30°, 45°, 60°, 75°). The schematic illustration is shown in the inset. (d) The corresponding histogram of the intensity of the peak at 613 cm$^{-1}$. (e–j) Electric field distribution by COMSOL simulation of an optical incidence angle from 0° to 75° for every 5°.

## 4. Conclusions

To summarize, we developed a flexible 3D Ag NPs/PVA cavity array composite structure that produces dense hot spots through the coupling effect between the focusing field of the incident light and the plasmon effect of Ag NPs in hemispherical cavities. It was shown that the corresponding morphologies of cavity layer structures with various concentrations of PVA solution. The effect of different sputtering times on the Raman properties of the substrate was discussed. Through theoretical simulation, it was concluded that under the excitation of the same cavity mode, Ag NP-cavity could produce a stronger coupling effect than intact Ag film-cavity, which is consistent with the experimental results. Using R6G, CV and MG as probe molecules for Raman detection, the composite substrate shows excellent enhancement effect, outstanding uniformity. Moreover, the signal stability of the substrate is evaluated by changing the incident angle of the laser and the bending times of the film. These results support that it is expected to be used for quantitative and qualitative SERS analysis in environmental monitoring and food safety.

**Supplementary Materials:** The following are available online at https://www.mdpi.com/article/10.3390/nano11123196/s1, Figure S1: UV–vis spectra of Ag NPs/PVA nanocavity (a) and PVA film (b) in a wavelength range between 200 nm and 800 nm. Figure S2: The Raman spectra of $10^{-10}$ M R6G molecules were detected on Ag NP-on PC and $10^{-2}$ M R6G molecules on PVA.

**Author Contributions:** Conceptualization, C.Z. and B.M.; methodology, X.Z.; software, G.G.; validation, C.L. and X.Z.; formal analysis, J.X.; investigation, J.Y.; data curation, Z.L.; writing—original draft preparation, J.L.; writing—review and editing, J.X. and C.Z.; project administration, S.X.; funding acquisition, B.M., C.Z., J.Y. and Z.L. All authors have read and agreed to the published version of the manuscript.

**Funding:** This research was funded by the National Natural Science Foundation of China (12174229, 11804200, 11974222, 12004226, 11904214, 11774208), Taishan Scholars Program of Shandong Province (tsqn201812104), Qingchuang Science and Technology Plan of Shandon Qingchuang Science and Technology Plan of Shandong Province (2019KJJ014, 2019KJJ017), a Project of Shandong Province Higher Educational Science and Technology Program (J18KZ011) and China Postdoctoral Science Foundation (2019M662423).

**Institutional Review Board Statement:** Not applicable.

**Informed Consent Statement:** Not applicable.

**Data Availability Statement:** Data are available within this manuscript and the Supplementary Material.

**Conflicts of Interest:** The authors declare no conflict of interest. The funders had no role in the design of the study; in the collection, analyses, or interpretation of data; in the writing of the manuscript, or in the decision to publish the results.

## References

1. Nong, J.P.; Tang, L.L.; Lan, G.L.; Luo, P.; Li, Z.C.; Huang, D.; Yi, J.; Wei, W. Enhanced Graphene Plasmonic Mode Energy for Highly Sensitive Molecular Fingerprint Retrieval. *Laser Photonics Rev.* **2021**, *15*, 2000300. [CrossRef]
2. Ding, Q.Q.; Wang, J.; Chen, X.Y.; Liu, H.; Li, Q.J.; Wang, Y.L.; Yang, S.K. Quantitative and Sensitive SERS Platform with Analyte Enrichment and Filtration Function. *Nano Lett.* **2020**, *20*, 7304–7312. [CrossRef]
3. Karthick Kannan, P.; Shankar, P.; Blackman, C.; Chung, C.H. Recent Advances in 2D Inorganic Nanomaterials for SERS Sensing. *Adv. Mater.* **2019**, *31*, 1803432. [CrossRef]
4. Li, C.H.; Xu, S.C.; Yu, J.; Li, Z.; Li, W.F.; Wang, J.H.; Liu, A.H.; Man, B.Y.; Yang, S.K.; Zhang, C. Local hot charge density regulation: Vibration-free pyroelectric nanogenerator for effectively enhancing catalysis and in-situ surface enhanced Raman scattering monitoring. *Nano Energy* **2021**, *81*, 105585. [CrossRef]
5. Zhang, C.; Li, C.H.; Yu, J.; Jiang, S.Z.; Xu, S.C.; Yang, C.; Liu, Y.J.; Gao, X.G.; Liu, A.H.; Man, B.Y. SERS activated platform with three-dimensional hot spots and tunable nanometer gap. *Sens. Actuators B Chem.* **2018**, *258*, 163–171. [CrossRef]
6. Barbillon, G.; Graniel, O.; Bechelany, M. Assembled Au/ZnO Nano-Urchins for SERS Sensing of the Pesticide Thiram. *Nanomaterials* **2021**, *11*, 2174. [CrossRef]
7. Tao, W.Y.; Zhao, A.W.; Sun, H.H.; Gan, Z.B.; Zhang, M.F.; Li, D.; Gao, H.Y. Periodic silver nanodishes as sensitive and reproducible surface-enhanced Raman scattering substrates. *RSC Adv.* **2014**, *4*, 3487–3493. [CrossRef]
8. Zhu, Y.Q.; Li, M.Q.; Yu, D.Y.; Yang, L.B. A novel paper rag as 'D-SERS' substrate for detection of pesticide residues at various peels. *Talanta* **2014**, *128*, 117–124. [CrossRef] [PubMed]
9. Zhao, X.F.; Liu, C.D.; Yu, J.; Li, Z.; Li, C.H.; Xu, S.C.; Li, W.F.; Man, B.Y.; Zhang, C. Hydrophobic multiscale cavities for high-performance and self-cleaning surface-enhanced Raman spectroscopy (SERS) sensing. *Nanophotonics* **2020**, *9*, 4761–4773. [CrossRef]

10. Lee, H.K.; Lee, Y.H.; Koh, C.S.L.; Phan-Quang, G.C.; Han, X.; Lay, C.L.; Sim, H.Y.F.; Kao, Y.C.; An, Q.; Ling, X.Y. Designing surface-enhanced Raman scattering (SERS) platforms beyond hotspot engineering: Emerging opportunities in analyte manipulations and hybrid materials. *Chem. Soc. Rev.* **2018**, *48*, 731–756. [CrossRef] [PubMed]
11. Ding, S.Y.; Yi, J.; Li, J.F.; Ren, B.; Wu, D.Y.; Panneerselvam, R.; Tian, Z.Q. Nanostructure-based plasmon-enhanced Raman spectroscopy for surface analysis of materials. *Nat. Rev. Mater.* **2016**, *1*, 16021. [CrossRef]
12. Yang, L.K.; Ren, Z.F.; Zhang, M.; Song, Y.L.; Li, P.; Qiu, Y.; Deng, P.Y.; Li, Z.P. Three-dimensional porous SERS powder for sensitive liquid and gas detections fabricated by engineering dense "hot spots" on silica aerogel. *Nanoscale Adv.* **2021**, *3*, 1012–1018. [CrossRef]
13. Dai, Z.G.; Xiao, X.H.; Wu, W.; Zhang, Y.P.; Liao, L.; Guo, S.S.; Ying, J.J.; Shan, C.X.; Sun, M.T.; Jiang, C.Z. Plasmon-driven reaction controlled by the number of graphene layers and localized surface plasmon distribution during optical excitation. *Light Sci. Appl.* **2015**, *4*, 342. [CrossRef]
14. Nong, J.P.; Tang, L.; Lan, G.L.; Luo, P.; Li, Z.C.; Huang, D.P.; Shen, J.; Wei, W. Combined visible plasmons of Ag nanoparticles and infrared plasmons of graphene nanoribbons for high performance surface-enhanced Raman and infrared spectroscopies. *Small* **2021**, *17*, 2004640. [CrossRef]
15. Lin, X.; Fang, G.Q.; Liu, Y.L.; He, Y.Y.; Wang, L.; Dong, B. Marangoni effect-driven transfer and compression at three-phase interfaces for highly reproducible nanoparticle monolayers. *J. Phys. Chem. Lett.* **2020**, *11*, 3573–3581. [CrossRef]
16. Dong, S.L.; Zhang, X.L.; Li, Q.; Liu, C.D.; Ye, T.; Liu, J.C.; Xu, H.; Zhang, X.G.; Liu, J.; Jiang, C.Z.; et al. Springtail-inspired Superamphiphobic Ordered Nanohoodoo Arrays with Quasi-doubly Reentrant Structures. *Small* **2020**, *16*, 2000779. [CrossRef]
17. Nong, J.P.; Tang, L.L.; Lan, G.L.; Luo, P.; Guo, C.C.; Yi, J.M.; Wei, W. Wideband tunable perfect absorptions of graphene plasmons via attenuated total reflections in Otto prism configurations. *Nanophotonics* **2020**, *9*, 645–655. [CrossRef]
18. Xiu, X.; Hou, L.; Yu, J.; Jiang, S.Z.; Li, C.H.; Zhao, X.F.; Peng, Q.Q.; Qiu, S.; Zhang, C.; Man, B.Y.; et al. Manipulating the surface-enhanced Raman spectroscopy (SERS) activity and plasmon-driven catalytic efficiency by the control of Ag NP/graphene layers under optical excitation. *Nanophotonics* **2021**, *10*, 1529–1540. [CrossRef]
19. Yu, J.; Guo, Y.; Wang, H.; Su, S.; Zhang, C.; Man, B.Y.; Lei, F.C. Quasi optical cavity of hierarchical ZnO nanosheets@Ag nanoravines with synergy of near- and far-field effects for in situ Raman detection. *J. Phys. Chem. Lett.* **2019**, *10*, 3676–3680. [CrossRef] [PubMed]
20. Liang, H.Y.; Li, Z.P.; Wang, W.Z.; Wu, Y.S.; Xu, H.X. Highly Surface-roughened "Flower-like" Silver Nanoparticles for Extremely Sensitive Substrates of Surface-enhanced Raman Scattering. *Adv. Mater.* **2009**, *21*, 4614–4618. [CrossRef]
21. Zhang, C.; Jiang, S.Z.; Huo, Y.Y.; Liu, A.H.; Xu, S.C.; Liu, X.Y.; Sun, Z.C.; Xu, Y.Y.; Li, Z.; Man, B.Y. SERS detection of R6G based on a novel graphene oxide/silver nanoparticles/silicon pyramid arrays structure. *Opt. Express* **2015**, *23*, 24811–24821. [CrossRef]
22. Graniel, O.; Iatsunskyi, I.; Coy, E.; Humbert, C.; Barbillon, G.; Michel, T.; Maurin, D.; Balme, S.; Miele, P.; Bechelany, M. Au-covered hollow urchin-like ZnO nanostructures for surface-enhanced Raman scattering sensing. *J. Mater. Chem. C* **2019**, *7*, 15066–15073. [CrossRef]
23. Liu, C.Y.; Xu, X.H.; Wang, C.D.; Qiu, G.Y.; Ye, W.C.; Li, Y.M.; Wang, D.Q. ZnO/Ag nanorods as a prominent SERS substrate contributed by synergistic charge transfer effect for simultaneous detection of oral antidiabetic drugs pioglitazone and phenformin. *Sens. Actuators B* **2020**, *307*, 127634. [CrossRef]
24. Yao, J.C.; Quan, Y.N.; Gao, M.; Gao, R.X.; Chen, L.; Liu, Y.; Lang, J.H.; Shen, H.; Zhang, Y.J.; Yang, L.L.; et al. AgNPs decorated Mg-doped ZnO heterostructure with dramatic SERS activity for trace detection of food contaminants. *J. Mater. Chem. C* **2019**, *7*, 8199–8208. [CrossRef]
25. Yu, J.; Yang, M.S.; Li, Z.; Liu, C.D.; Wei, Y.S.; Zhang, C.; Man, B.Y.; Lei, F.C. Hierarchical Particle-In-Quasicavity Architecture for Ultratrace in Situ Raman Sensing and Its Application in Real-Time Monitoring of Toxic Pollutants. *Anal. Chem.* **2020**, *92*, 14754–14761. [CrossRef]
26. Chen, Y.C.; Sun, M.T. Two-dimensional WS$_2$/MoS$_2$ heterostructures: Properties and applications. *Nanoscale* **2021**, *13*, 5594–5619. [CrossRef] [PubMed]
27. Dai, F.; Horrer, A.; Adam, P.M.; Fleischer, M. Accessing the Hotspots of Cavity Plasmon Modes in Vertical Metal–Insulator–Metal Structures for Surface Enhanced Raman Scattering. *Adv. Opt. Mater.* **2020**, *8*, 1901734. [CrossRef]
28. Guerra Hernández, L.A.; Huidobro, P.A.; Cortés, E.; Maier, S.A.; Fainstein, A. Resonant Far- to Near-Field Channeling in Synergetic Multiscale Antennas. *ACS Photonics* **2019**, *6*, 1466–1473. [CrossRef]
29. Li, W.; Xiong, L.; Li, N.; Pang, S.; Xu, G.; Yi, C.; Wang, Z.; Gu, G.; Li, K.; Li, W.; et al. Tunable 3D light trapping architectures based on self-assembled SnSe$_2$ nanoplate arrays for ultrasensitive SERS detection. *J. Mater. Chem.* **2019**, *7*, 10179–10186. [CrossRef]
30. Li, Y.; Shang, Y.; Lin, J.; Li, A.; Wang, X.; Li, B.; Guo, L. Temperature-Induced Stacking to Create Cu$_2$O Concave Sphere for Light Trapping Capable of Ultrasensitive Single-Particle Surface-Enhanced Raman Scattering. *Adv. Funct. Mater.* **2018**, *28*, 1801868. [CrossRef]
31. Liu, B.; Yao, X.; Chen, S.; Lin, H.; Yang, Z.; Liu, S.; Ren, B. Large-Area Hybrid Plasmonic Optical Cavity (HPOC) Substrates for Surface-Enhanced Raman Spectroscopy. *Adv. Funct. Mater.* **2018**, *28*, 1802263. [CrossRef]
32. Liu, Y.; Tian, X.; Guo, W.; Wang, W.; Guan, Z.; Xu, H. Real-time Raman detection by the cavity mode enhanced Raman scattering. *Nano Res.* **2019**, *12*, 1643–1649. [CrossRef]
33. de Aberasturi, D.J.; Henriksen-Lacey, M.; Litti, L.; Langer, J.; Liz-Marzán, L.M. Using SERS Tags to Image the Three-Dimensional Structure of Complex Cell Models. *Adv. Funct. Mater.* **2020**, *30*, 1909655. [CrossRef]

34. Mao, P.; Liu, C.X.; Chen, Q.; Han, M.; Maier, S.A.; Zhang, S. Broadband SERS detection with disordered plasmonic hybrid aggregates. *Nanoscale* **2020**, *12*, 93–102. [CrossRef] [PubMed]
35. Park, H.J.; Cho, S.; Kim, M.; Jung, Y.S. Carboxylic acid-functionalized, graphitic layer-coated three-dimensional SERS substrate for label-free analysis of Alzheimer's disease biomarkers. *Nano Lett.* **2020**, *20*, 2576–2584. [CrossRef] [PubMed]
36. Yang, H.; Gun, X.Y.; Pang, G.H.; Zheng, Z.X.; Li, C.B.; Yang, C.; Wang, M.; Xu, K.C. Femtosecond laser patterned superhydrophobic/hydrophobic SERS sensors for rapid positioning ultratrace detection. *Opt. Express* **2021**, *29*, 16904–16913. [CrossRef] [PubMed]
37. Zuo, Z.W.; Sun, L.Y.; Guo, Y.B.; Zhang, L.J.; Li, J.H.; Li, K.G.; Cui, G.L. Multiple plasmon couplings in 3D hybrid Au-nanoparticles-decorated Ag nanocone arrays boosting highly sensitive surface enhanced Raman scattering. *Nano Res.* **2021**, *15*, 317–325. [CrossRef]
38. Zuo, Z.; Zhang, S.; Wang, Y.; Guo, Y.; Sun, L.; Li, K.; Cui, G. Effective plasmon coupling in conical cavities for sensitive surface enhanced Raman scattering with quantitative analysis ability. *Nanoscale* **2019**, *11*, 17913–17919. [CrossRef]
39. Zhu, C.; Zhao, Q.; Meng, G.; Wang, X.; Hu, X.; Han, F.; Lei, Y. Silver nanoparticle-assembled micro-bowl arrays for sensitive SERS detection of pesticide residue. *Nanotechnology* **2020**, *31*, 205303. [CrossRef]
40. Lee, Y.; Lee, J.; Lee, T.K.; Park, J.; Ha, M.; Kwak, S.K.; Ko, H. Particle-on-Film Gap Plasmons on Antireflective ZnO Nanocone Arrays for Molecular-Level Surface-Enhanced Raman Scattering Sensors. *ACS Appl. Mater. Interfaces* **2015**, *7*, 26421–26429. [CrossRef]
41. Mao, P.; Liu, C.; Favraud, G.; Chen, Q.; Han, M.; Fratalocchi, A.; Zhang, S. Broadband single molecule SERS detection designed by warped optical spaces. *Nat. Commun.* **2018**, *9*, 5428. [CrossRef] [PubMed]
42. Xu, K.C.; Zhou, R.; Takei, K.; Hong, M.H. Toward Flexible Surface-Enhanced Raman Scattering (SERS) Sensors for Point-of-Care Diagnostics. *Adv. Sci.* **2019**, *6*, 1900925. [CrossRef] [PubMed]
43. Pandey, P.; Vongphachanh, S.; Yoon, J.; Kim, B.; Choi, C.; Sohn, J.I.; Hong, W.K. Silver nanowire-network-film-coated soft substrates with wrinkled surfaces for use as stretchable surface enhanced Raman scattering sensors. *J. Alloys Compd.* **2021**, *859*, 157862. [CrossRef]
44. Yi, Z.; Niu, G.; Luo, J.; Kang, X.; Yao, W.; Zhang, W.; Yi, Y.; Yi, Y.; Ye, X.; Duan, T.; et al. Ordered array of Ag semishells on different diameter monolayer polystyrene colloidal crystals: An ultrasensitive and reproducible SERS substrate. *Sci. Rep.* **2016**, *6*, 32314. [CrossRef]
45. Qin, F.F.; Su, M.; Zhao, J.L.; Moqaddam, A.M.; Del Carro, L.; Brunschwiler, T.; Kang, Q.J.; Song, Y.L.; Derome, D.; Carmeliet, J. Controlled 3D nanoparticle deposition by drying of colloidal suspension in designed thin micro-porous architectures. *Int. J. Heat Mass Transf.* **2020**, *158*, 120000. [CrossRef]
46. Hess, O.; Pendry, J.B.; Maier, S.A.; Oulton, R.F.; Hamm, J.M.; Tsakmakidis, K.L. Active nanoplasmonic metamaterials. *Nat. Mater.* **2012**, *11*, 573–584. [CrossRef]
47. Galinski, H.; Favraud, G.; Dong, H.; Gongora, J.S.T.; Favaro, G.; Dobeli, M.; Spolenak, R.; Fratalocchi, A.; Capasso, F. Scalable, ultra-resistant structural colors based on network metamaterials. *Light Sci. Appl.* **2017**, *6*, 16233. [CrossRef]
48. Nie, S.; Emory, S. Probing single molecules and single nanoparticles by surface-enhanced Raman scattering. *Science* **1997**, *275*, 1102–1106. [CrossRef]
49. Li, X.H.; Choy, W.C.H.; Ren, X.G.; Zhang, D.; Lu, H.F. Highly Intensified Surface Enhanced Raman Scattering by Using Monolayer Graphene as the Nanospacer of Metal Film-Metal Nanoparticle Coupling System. *Adv. Funct. Mater.* **2014**, *24*, 3114–3122. [CrossRef]

Article

# Optical Properties of Ag Nanoparticle Arrays: Near-Field Enhancement and Photo-Thermal Temperature Distribution

Daobin Luo [1,*], Pengcheng Hong [1], Chao Wu [2], Shengbo Wu [1] and Xiaojing Liu [1]

[1] School of Arts and Sciences, Shaanxi University of Science & Technology, Xi'an 710021, China
[2] Xian Institute of Space Radio Technology, Xi'an 710000, China
* Correspondence: luodaobin@sust.edu.cn

**Abstract:** The near-field and photo-thermal properties of nanostructures have always been the focus of attention due to their wide applications in nanomaterials. In this work, we numerically investigate the near-field and photo-thermal temperature distribution in a nanoparticle array when the scattering light field among particles is considered. 'Hot spots', which represent strong electric field enhancement, were analyzed at the difference of the particle size, particle spacing and the polarization direction of the incident light. Interestingly, it is found that the position of the 'hot spots' does not rotate with the polarization direction of the incident light and always remains in the particle gaps along the line between particle centers. Moreover, the near-field is independent of the polarization in some special areas, and the factor of near-field enhancement keeps constant in these spots when the illumination polarization varies. As for photo-induced heating, our results show that both the temperature of the structure center and maximum temperature increase linearly with the particle number of the array while decreasing with the increase in particle spacing. This work provides some theoretical considerations for the near-field manipulation and photo-thermal applications of nanoarrays.

**Keywords:** Ag nanoparticles; square arrays; near-field enhancement; photo-induced heating; photothermal temperature distribution

## 1. Introduction

Noble metal nanoparticles have attracted extensive interest in physics, chemistry, and biology due to their surface plasmon properties. When the free charges oscillate, collectively driven by the incident light at resonance frequency, the local surface plasmon (LSP) phenomenon occurs [1]. Noble metal nanoparticles have optical properties of strong light absorption and near-field enhancement in the visible band when LSP occurs. The LSP of noble metal nanoparticles has been widely studied in the last 20 years [2]. Researchers have made great contributions to the optical properties of nanoparticles with different shapes [3] and of periodic array structures [4,5]. Our group has also studied the factor of near-field enhancement and determined the near-field distribution of Ag dimer [6]. The optical properties of nanoparticles have been widely applied in surface enhanced Raman scattering [7], photovoltaic [8,9], photocatalytic and other fields [10].

At the same time, the strong light absorption of noble metal nanoparticles in the visible light band increases the temperature of the nanoparticles themselves, and the photo-thermal effect is another important property of plasmonic structures [11]. For a long time in the past, the photo-thermal effect was regarded as a negative effect by researchers and was reduced or suppressed [12]. Later, researchers gradually found that the photo-thermal properties of nanoparticles could be applied to thermos-physics [13], biomedicine and other fields [14], and it has attracted increasing attention since [15]. Now, the photo-thermal effect of nanostructure has been widely used in the fields of photo-thermal therapy [16], drug release [17], solar energy collection [18] and other areas.

Due to the long-range ordered periodicity, nanoarrays have the characteristics of expanding the electromagnetic enhancement space, generating collective lattice resonances [4,5], increasing scattering efficiency, and enhancing thermal accumulation effects [19]. Thus, we investigated the absorption, scattering, near-field enhancement and photothermal properties of Ag nanoarrays. In this article, discrete dipole approximation (DDA) and thermal Green's function method are introduced to investigate near-field enhancement and photo-thermal temperature distribution, respectively. The influences of particle radius, particle spacing, particle number of the array and illumination polarization angle on the unique optical properties of Ag nanoarrays are discussed. The purpose of this work is to understand the near-field and photo-thermal temperature distribution of Ag nanoarrays and to explore a method for designing the nanostructure in photo-thermal and near-field regulation.

## 2. Theories and Methods

Figure 1 is a schematic diagram of a square array of spherical nanoparticles. $R$ and $d$ denote the particle radius and spacing, respectively. The plane of the array is located at the $xoy$ and the center of the array structure is located at the coordinate origin. The incident light travels along the negative direction of the z-axis.

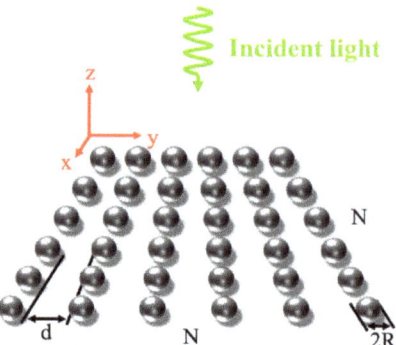

**Figure 1.** Schematic diagram of square array of spherical nanoparticles.

We used DDA to study the near-field enhancement, absorption, and scattering properties of the nanoparticle array. Based on the absorption properties of nanoparticles and thermal Green's function method, the photo-thermal temperature distribution was investigated.

*2.1. Discrete Dipole Approximation Method*

DDA is a method used to calculate the scattering and absorption of electromagnetic wave radiation by targets with an arbitrary geometry whose size is less than or equal to the wavelength of incident light [20]. The main idea is to treat the target model as a set of multiple point dipoles. Each point dipole occupies a position in the target lattice. They are polarized to generate an electric dipole moment when the incident electromagnetic wave illuminates the target. Then, each point dipole is affected not only by the external electric field but the scattering field of other surrounding point dipoles. Therefore, the electromagnetic field around the target point dipole is the superposition of the incident field and the scattering field of the other multiple point dipoles. Draine and Flatau [21] developed a ddscat7.3 program code based on DDA theory to calculate the electric field distribution of electromagnetic scattering from various nanostructures. In this article, ddscat7.3 is used to calculate the absorption, scattering, and near-field enhancement of the Ag nanoparticle array.

## 2.2. Thermal Green's Function Method

As for the plasma thermal behavior of metal nanoparticle assembly, we commonly use the thermoplasmonics theory [15]. The thermal effect produced by the interaction between incident light and particles increases the temperature of the particles themselves and the temperature of the surrounding medium increases due to thermal diffusion. As shown in Figure 1, the heat absorption power of the *ith* nanoparticle in the structure is:

$$Q_i = \frac{1}{2} n c \varepsilon_0 \sigma_{abs} |E_i^{ext}|^2, \tag{1}$$

where $n$ stands the refractive index of the surrounding medium, $c$ stands the speed of light, $\varepsilon_0$ stands the dielectric constant of vacuum, $E_i^{ext}$ stands the external electric field intensity around the particles after the incident light interacts with the nanoparticle array. The electric field intensity can be calculated using the DDA method. Siahpoush et al. [22] used this method in the study of the effect of plasmonic coupling on the photo-thermal behavior of random nanoparticles. $\sigma_{abs}$ stands the absorption cross-section of nanoparticles. Here, $\sigma_{abs} + \sigma_{sca} = \sigma_{ext}$. $\sigma_{sca}$ and $\sigma_{ext}$ stand the scattering and extinction cross-sections, respectively. Correspondingly, $Q_{abs}$, $Q_{sca}$ and $Q_{ext}$ stand the absorption efficiency, the scattering and extinction efficiencies, respectively. For spherical nanoparticles, $\sigma_{abs} = Q_{abs} \cdot \pi R^2 (\sigma_{sca} = Q_{sca} \cdot \pi R^2, \sigma_{ext} = Q_{ext} \cdot \pi R^2)$, which shows that $\sigma_{abs}$ is a quantity with dimensions of area [23].

The steady-state temperature distribution of the *ith* nanoparticle isolated in a homogeneous medium is considered. The details of the mechanism explanation and process are given in Reference [15]. For $|r - r_i| > R$,

$$T(r) = \frac{Q_i}{4\pi \kappa_s |r - r_i|} + T_\infty. \tag{2}$$

For $|r - r_i| \leq R$,

$$T(r) = \frac{Q_i}{4\pi \kappa_s R} + T_\infty. \tag{3}$$

where $\kappa_s$ stands the thermal conductivity of the surrounding medium, $T_\infty$ stands the temperature at which the distance is infinite (ambient temperature).

In the case of an array structure, the increased temperature at the position vector $r$ in the outer space of the particles is provided by the collection of other $j$ nanoparticles.

$$\Delta T(r) = \sum_{j=1}^{N^2} G(r, r_j) Q_j, \tag{4}$$

where

$$G(r, r_j) = \frac{1}{4\pi \kappa_s |r - r_j|}. \tag{5}$$

The increased temperature inside the *ith* particle is

$$\Delta T_i = \sum_{\substack{j=1 \\ j \neq i}}^{N^2} G(r_i, r_j) Q_j + \frac{Q_i}{4\pi \kappa_s R}. \tag{6}$$

The first term of the above formula is the temperature contribution of $N^2 - 1$ particles around *ith* particle. The second term is self-contribution of *ith* particle. Therefore, Equation (6) can also be written as

$$\Delta T_i = \Delta T_i^{ext} + \Delta T_i^{self}. \tag{7}$$

## 3. Results and Discussions

The optical properties of the Ag nanoparticle array were investigated carefully. In the following section, the influences of the radius, spacing and number of the particles and the illumination polarization angle on the absorption, near-field enhancement and photo-thermal temperature distribution are discussed in detail. The main parameters we set are as follows: The refractive index function file of Ag is given in Reference [24]. The refractive index of ambient medium is 1.33. The thermal conductivity of the medium is $0.599 W/(m \cdot K)$ (considered in the water environment). The irradiance of incident light is $1.27 \times 10^8 W/m^2$.

### 3.1. Optical Properties of Array Structure
#### 3.1.1. The Influence of Particle Spacing

In the case of y-polarized incident light, we calculated and analyzed the effects of particle spacing $d$ on the optical properties of the array with 3 × 3 particles. Figure 2 shows the relationship of absorption and scattering efficiencies versus the particle spacing. The results show that the absorption and scattering peaks were broadened and redshifted with the decrease of the particle spacing, especially when $d \leq 2R$. It is considered that the surface plasmon phenomenon occurs when light is incident upon the array. In our case, each nanoparticle can be simply regarded as point dipoles, and the electron cloud inside them produces reciprocating oscillation behavior under the action of the electric field from incident light, as shown in Figure 3. While the nanoparticle spacing is close, the free electron oscillation inside each particle has different driven external field due to the scattering field from the surrounding particles, leading to the more oscillation frequencies of the free electronics. Multi-frequency of electron oscillation causes a broadened resonance peak. Considering the charge distribution of the particles, a decrease in the particle spacing weakens the restoring force of free electron oscillations inside particles due to the charge interaction of adjacent particles parallel to polarization (see Figure S1). Moreover, the resonance frequency decreases, and the resonance peak red-shifts. The weakening of the restoring force leads to the decrease in the amplitude of the electron oscillation. The decrease in the amplitude denotes the decrease in the absorption, and absorption efficiency becomes smaller. Generally speaking, the scattering energy includes the contribution of the first scattering from the particle and the multiple scattering between particles. Comparing the absorption efficiency, the intension of the scattering efficiency changes little when the particle spacing changes. The main reason is the scattering contribution of the light passing particle's surface for the first time, not the multiple scattering between particles.

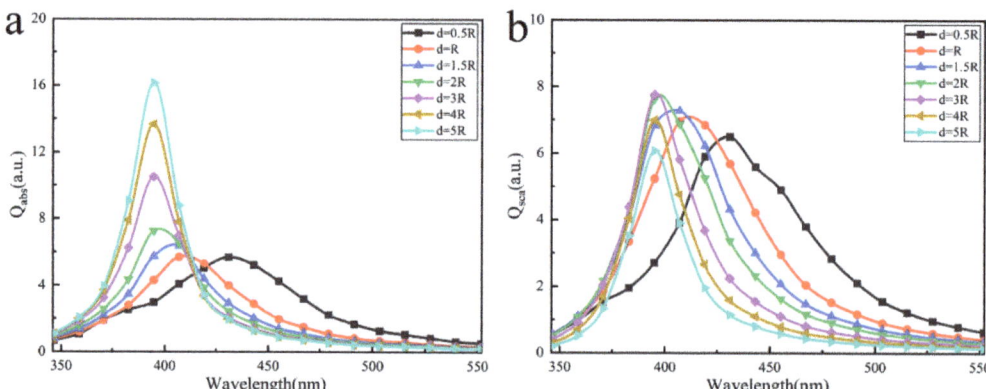

**Figure 2.** The absorption and scattering efficiencies of the 3 × 3 square array of Ag nanoparticles at different particle spacing with radius $R = 14 nm$: (**a**) absorption efficiency (**b**) scattering efficiency.

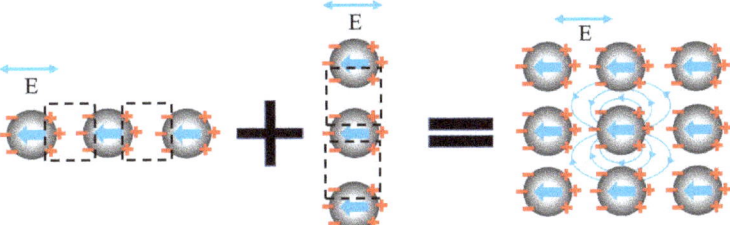

**Figure 3.** Schematic diagram of near-field coupling between particles in array under horizontal polarization.

Figure 4 shows the near-field distribution at different particle spacings. The results show that the 'hot spots'—strong electric field enhancement regions—appear in the particle gaps in the direction of polarization, especially when the particle spacing is smaller than the particle radius. 'Hot spots' become weaker gradually as the particle spacing increases, and electric field enhancement regions are limited to the light polarization direction near the surface of nanoparticles. 'Hot spots' exit because near-field coupling occurs between the nanoparticles when the particle spacing is small.

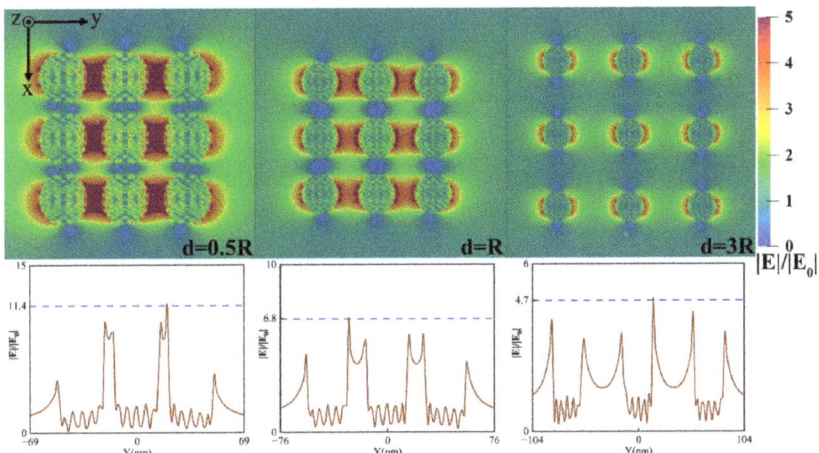

**Figure 4.** Near-field enhancement of the 3 × 3 square array of Ag nanoparticles with the radius of 14 nm and the incident light wavelength of 532nm: near-field distribution with different particle spacing (the top) and field enhancement factor curve along the y-axis of the array center (the bottom).

3.1.2. The Influence of Particle Radius

Similarly, in the case of y-polarized incident light, we calculated and analyzed the effects of particle radius $R$ on the optical properties. Figure 5 shows the relationship of absorption and scattering efficiencies versus the particle radius. The results show that the absorption and scattering peaks are broadened and redshifted with the increase of the particle radius. Especially when the radius is 23 nm and 26 nm, two absorption peaks appear in Figure 5a and three scattering peaks appear in Figure 5b. While the nanoparticles' radius increases, the driven electric field of incident light in the particle has a nonuniform phase and causes the phase difference of electron movement inside the particle, and 'Phase delay' effects occur. Moreover, each particle also has inconsistent oscillation frequencies due to the influence of the multiple scattering field. These lead to the broadening of the resonance peak. For the resonance peak shifting, considering the electric field force exerted by adjacent particles' charges (see Figure S2), the restoring force of free electron oscillations inside particles decreases as the number of surface free net charges of the neighboring particles increase. Therefore, the resonance frequency decreases and the resonance peak red-shifts as the particle radius increases. As we know, increasing particle radius leads to a decrease in absorption efficiency and an increase in scattering efficiency for a single particle (see Figure S3). The same occurred in nanoparticle arrays. On the one hand, the decrease in amplitude of the electron oscillation denotes the energy of electrons absorbing electromagnetic waves decreases. On the other hand, the contact area between particles and light increases with the particle's radius. Therefore, the absorption peak falls while the scattering peak rises when increasing the particle radius. For single particle, 2 × 2 and 3 × 3 arrays, we found that the peak number of scattering or absorption increases with particle number (see Figure 5 and Figure S3).

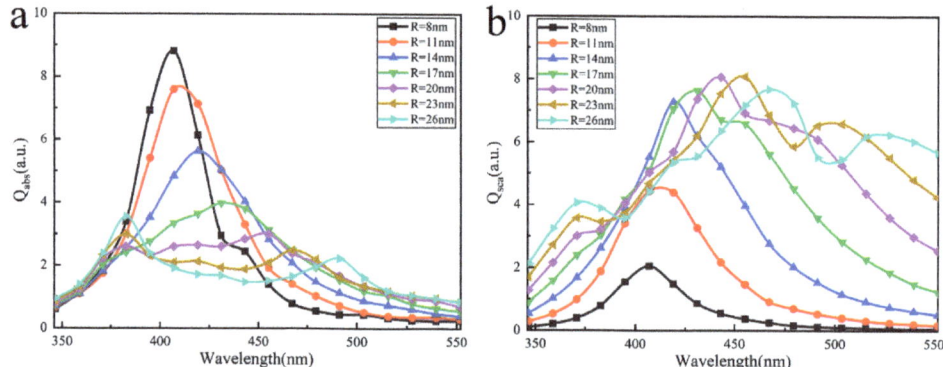

**Figure 5.** The absorption and scattering efficiencies of the 3 × 3 square array of Ag nanoparticles with different particle radiuses with spacing $d = 10nm$ : (**a**) absorption efficiency (**b**) scattering efficiency.

Figure 6 shows the near-field distribution when the particle radius varies. The result shows that 'hot spots' appear in the polarization direction of particle gaps with the increase in particle radius. Firstly, the absorption and scattering efficiencies of arrays with different radii are different at the wavelength of 532 nm, which affects the field or energy distribution around particles. Secondly, the number of free electrons in the particle goes up with the increase in particle radius, and more free electrons take apart in the oscillation due to drive by incident light, so the coupling between particles is significantly improved. However, different phenomena may occur at other wavelengths.

### 3.1.3. The Influence of Illumination Polarization

Furthermore, we investigated the influences of different illumination polarization on the optical properties of the array with 3 × 3 particles. Figure 7 shows the absorption and scattering efficiencies spectra of the Ag nanoparticle array. The result shows that the absorption and scattering efficiencies of the structure remain entirely unchanged as the polarization angle of the incident light increases from 0° to 90°. Firstly, the sphere is a highly centrosymmetric structure, and Ag nanoparticles in the array are isotropic and uniform materials. Therefore, the dependence of the polarization angle only needs to be considered as the particle arrangement position. Secondly, the square array conforms to orthogonal symmetry. When polarized light at any angle illuminates the array, polarization orthogonally decomposed into transverse and longitudinal modes [25]. To sum up the above two points, it is easy to conclude that the transverse and longitudinal modes of the absorption and scattering efficiencies are equal in the square array. Therefore, the absorption and scattering efficiencies are naturally independent of the incident polarization.

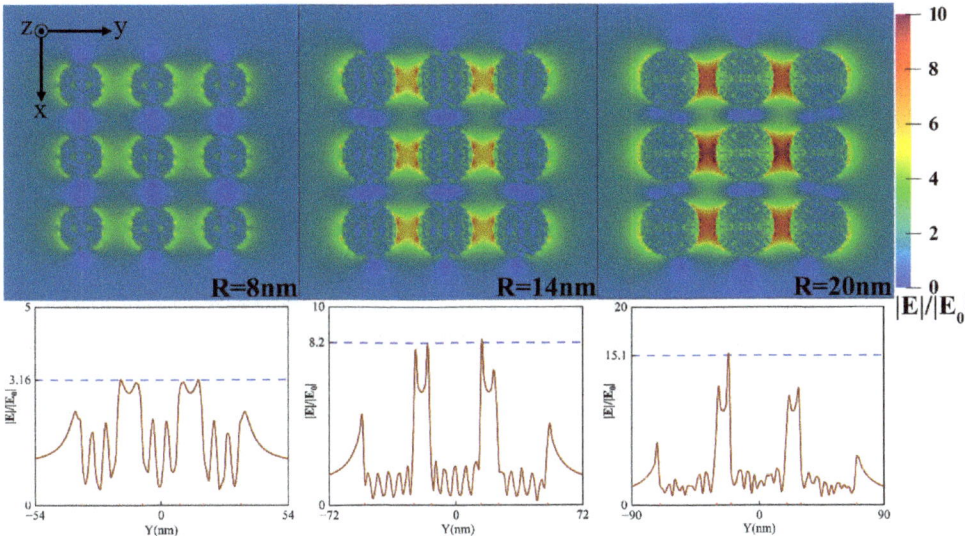

**Figure 6.** Near-field enhancement of the 3 × 3 square array of Ag nanoparticles with the particle spacing of 10nm and the incident light wavelength of 532 nm: near-field distribution with different radius (the top) and field enhancement factor curve along the y-axis of the array center (the bottom).

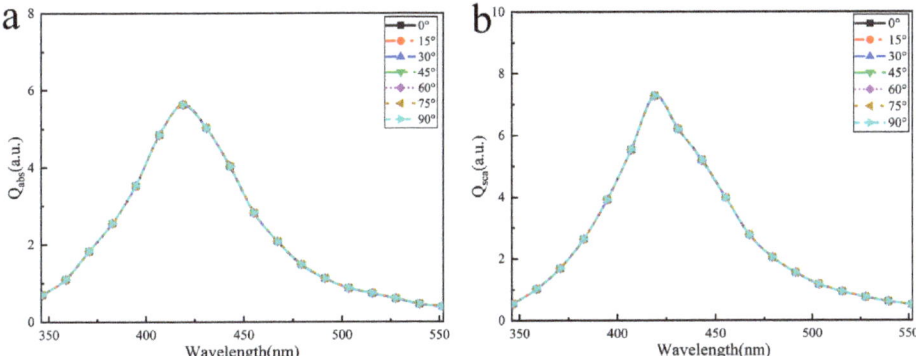

**Figure 7.** The absorption and scattering efficiencies of the 3 × 3 square array of Ag nanoparticles at different illumination polarization angles with particle radius $R = 14nm$ and particle spacing $d = 10nm$: (**a**) absorption efficiency (**b**) scattering efficiency.

Then, we investigated the near-field enhancement effect at different illumination polarization angles from 0° to 45°. The near-field enhancement of the single Ag particle, 2 × 2 and 3 × 3 arrays of nanoparticles were calculated, shown in Figure 8. It is easy to observe that the direction of the near-field enhancement region rotates with the illumination polarization angle and the rotation is completely synchronous for a single nanoparticle. For arrays, it is noted that 'hot spots' tend to remain in the gap between nearest neighboring particles along the y-direction (or x-direction), although their intensity varies with the deflection of the polarization direction.

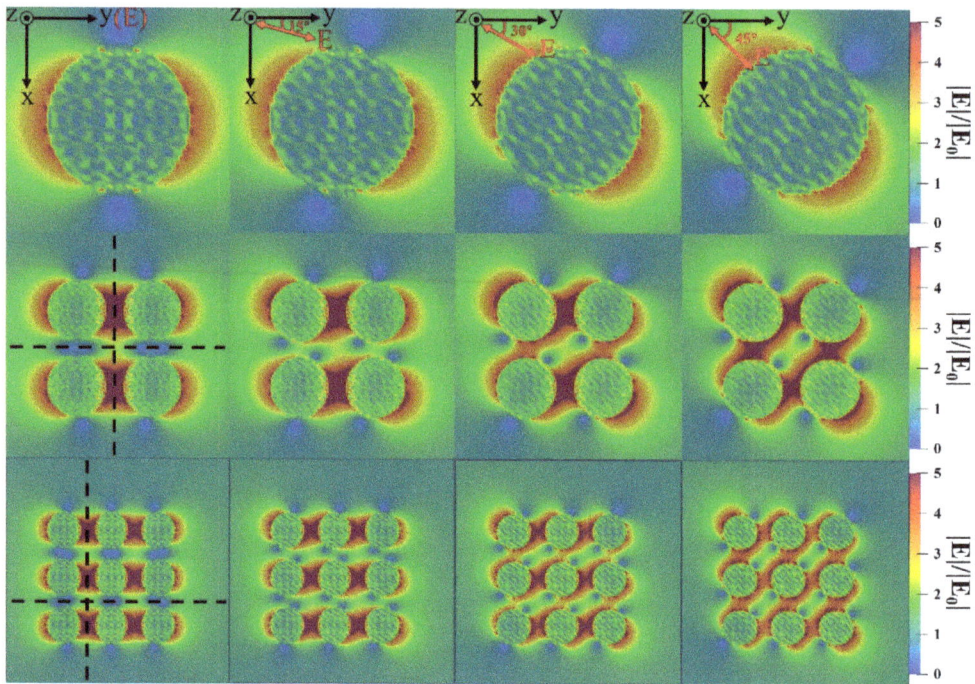

**Figure 8.** Near-field distribution of single and square arrays at different illumination polarization angles with the incident light wavelength of 532nm: the single ($R = 20nm$), $2 \times 2$ array ($R = 20nm$, $d = 10nm$) and $3 \times 3$ array ($R = 14nm$, $d = 10nm$).

Interestingly, we found that the near-field intensity of the center of any four adjacent particles is kept constant when the illumination polarization changes. We call these points 'constant spots'. Figure 9 shows the field enhancement factors along the dotted lines in Figure 8. We found that near-field enhancement remains invariable in 'constant spots' when the polarization changes, whether with $2 \times 2$ or $3 \times 3$ arrays. However, 'hot spots' arrive the strongest at the polarization direction of y-direction or x-direction, and monotonically decrease with the increase of the illumination polarization angle. The main reason is that the electron cloud oscillation inside the nanoparticles varies with illumination polarization. When the illumination polarization angle is located at the x-axis or y-axis, the distance of the surface oscillation charges between particles' surfaces arrives at a minimum, and the near-field enhancement reaches a maximum in the gap among particles due to the superposition of electric field excited by surface charges of neighbor particles. Interestingly, some researchers reported polarization independent of the photo-thermal temperature or absorption in nanostructures [25,26]. Here, we found polarization shielding or polarization independent of near-field enhancement in nanoparticle arrays. Our findings may contribute to the study of the near-field invariance of nano structures and provide new considerations for near-field modulation and control.

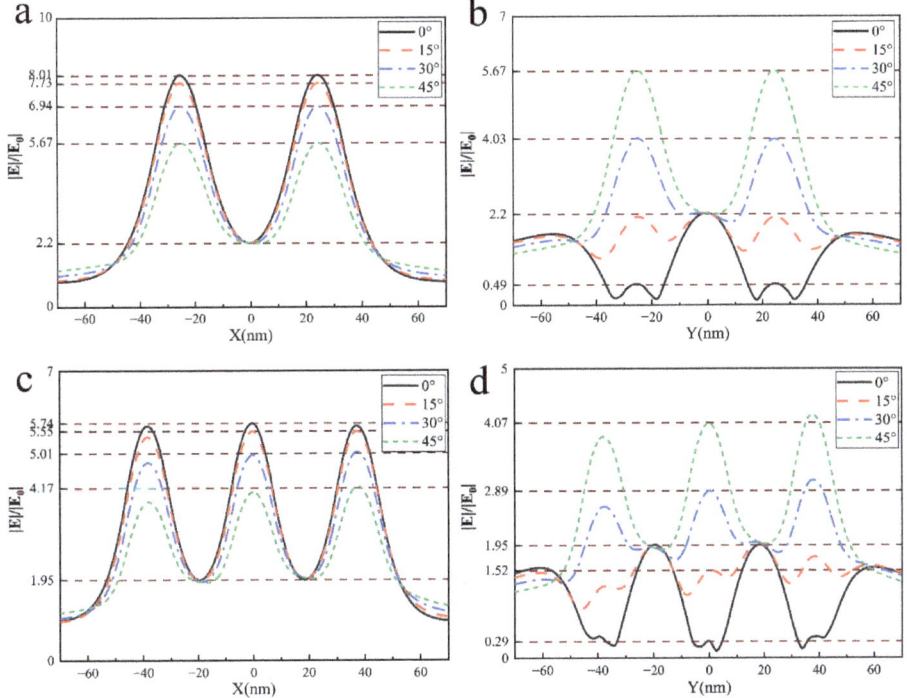

**Figure 9.** Enhancement factor curve at the dotted line in Figure 8 (**a**) x-direction of 2 × 2 array, (**b**) y-direction of 2 × 2 array (**c**) x-direction of 3 × 3 array, (**d**) y-direction of 3 × 3 array.

*3.2. Photo-Induced Heating of Array Structure*

In this section, the thermal Green's function method is used to investigate the photo-induced heating of spherical Ag nanoparticle arrays. To discuss the central temperature and temperature state of the array, $\zeta$, a dimensionless parameter to indicate the state of temperature is defined as [11]

$$\zeta = \frac{\Delta T_0^{self}}{\Delta T_0^{ext}}, \tag{8}$$

where $\Delta T_0^{self}$ stands the temperature contribution of the central particle itself, and $\Delta T_0^{ext}$ stands the temperature contribution of the surrounding $N^2 - 1$ nanoparticles. If $\zeta \gg 1$, indicates that self-contribution is dominant, and the structure is a temperature confinement regime. If $\zeta \ll 1$, it indicates that external contribution is dominant, and the structure is a temperature delocalization regime. When multiple scattering is negligible, the external field $E_i^{ext}$ is equal to the illuminating incident field $E_i^{inc}$. Based on the reference [11], we draw the expression without considering the multiple scattering effect between particles as

$$\zeta = \frac{d + 2R}{3(N - 1)R}. \tag{9}$$

We used the thermal Green's function method to calculate the dimensionless parameter under multiple scattering and compared with Equation (9) (see Figure 10) and explored this dimensionless parameter of the array varying with the radius, particle spacing and particle number. From the calculation results in Figure 10, it can be seen that the state parameters with multiple scattering are between 0.02 and 1.04, which indicates that the structure under consideration is basically in the temperature delocalization state. It is obvious that the dimensionless parameter decreases with the increase of radius and particle number, shown in Figure 10a,c. Interestingly, the parameters increase linearly with particle spacing.

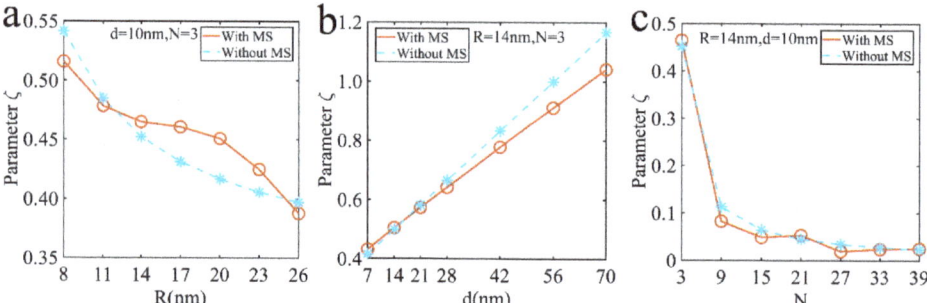

**Figure 10.** The relationship between the dimensionless parameter $\zeta$ and the parameter of the array structures: (**a**) radius, (**b**) particle spacing, (**c**) particle number.

In order to observe the temperature distribution of the array more intuitively, we calculated the temperature distribution of some structures and the temperature variation curve of the horizontal center line, shown in Figure 11. Each particle can be regarded as an individual nano heat source when $\zeta = 1.04$. However, as $\zeta$ decreases, the thermal coupling effect becomes obvious, and the temperature of the array tends to be more uniform. So, we can draw the conclusion that the parameter $\zeta$ of our structures denotes the temperature delocalization regime.

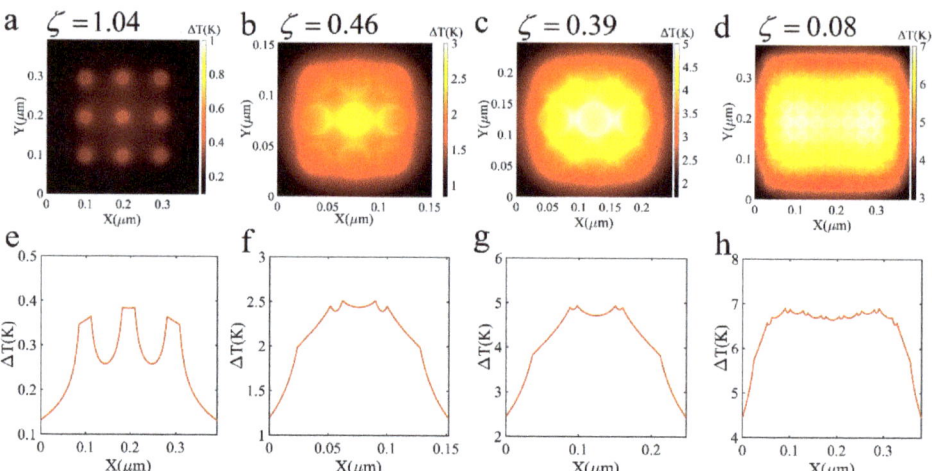

**Figure 11.** Photo-induced heating of Ag nanoparticle array illuminated by uniform 532nm light with light irradiance $I = 1.27 \times 10^8 W/m^2$: (**a**) $3 \times 3$ array ($R = 14nm$, $d = 70nm$, $Q_{abs} = 0.2635$), (**b**) $3 \times 3$ array ($R = 14nm$, $d = 10nm$, $Q_{abs} = 0.5601$), (**c**) $3 \times 3$ array ($R = 26nm$, $d = 10nm$, $Q_{abs} = 1.0154$), (**d**) $9 \times 9$ array ($R = 14nm$, $d = 10nm$, $Q_{abs} = 0.5601$), (**e–h**) Temperature variation curve of horizontal center line.

Similarly, to analyze the temperature distribution in detail, we explored the center and maximum temperature of the array varying with the radius, particle spacing and particle number. We found that the maximum temperature is not the position of the array center, which is reflected in Figure 11. As for maximum and center temperatures, they significantly rise quickly as the particle radius increases when $8nm \leq R \leq 20nm$, shown in

Figure 12a. However, the effect of plasmonic coupling leads to the decrease in temperature when $R > 20 nm$. The center and maximum temperature of the array decrease when the particle spacing increases, and it increases linearly with the number of particles, as shown in Figure 12b,c.

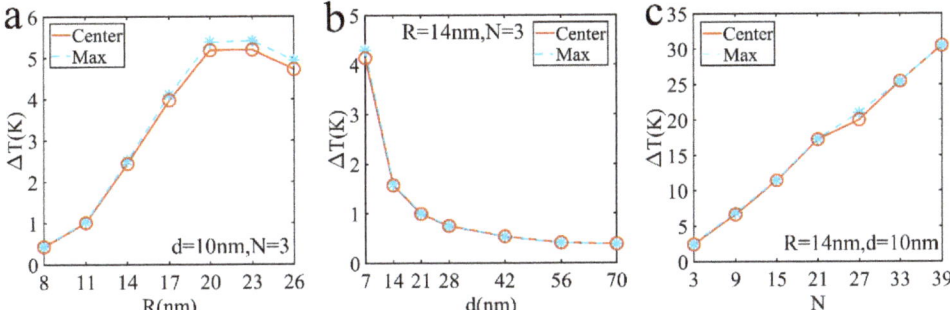

**Figure 12.** The central and maximum temperature of Ag nanoparticle array structure: (**a**) radius dependent, (**b**) particle spacing dependent, (**c**) particle number dependent.

It is always a major subject to control the temperature distribution by using nanoscale particles, and it has been reported that specific temperature distribution can be achieved by designing structures [27]. This can be used to manipulate single cell adhesion [28] and other fields [29]. It has broad application prospects. In the study, we found that, as shown in Figure 13, the plasma coupling effect led to an unconventional temperature distribution in the array, which varied with the number of particles. Therefore, we consider that adjusting the plasma coupling mode can also be a way to achieve temperature distribution regulation.

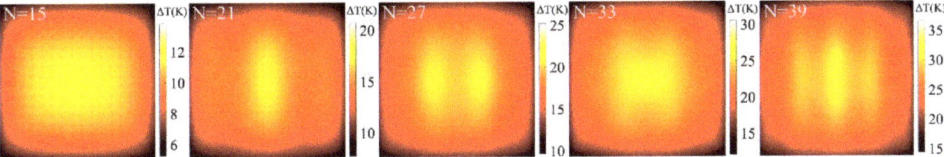

**Figure 13.** Temperature distribution under the effect of plasmonic coupling.

## 4. Conclusions

The near-field enhancement and photo-thermal temperature distribution of Ag nanoparticle arrays were investigated by using DDA and thermal Green's function methods. The results show that in the polarization direction, 'hot spots' appear in the particle gap when the particle spacing is smaller than the particle radius. The particle radius also affects the generation of 'hot spots'. It is noted that the absorption and scattering efficiencies of the array are unchanged when the illumination polarization angle is changed, and the position of the 'hot spots' do not rotate with the polarization direction of the illumination light. Moreover, we find that 'constant spots', which are located at the array center composed of four adjacent particles, and the factor of near-field enhancement stay constant in these spots when the illumination polarization varies. As for photo-induced heating, we determined a parameter $\zeta$ to describe the temperature delocalization state. Each particle can be regarded as an individual nano-heat source when the dimensionless parameter $\zeta = 1.04$. However, as $\zeta$ decreases, the thermal coupling effect becomes obvious, and the temperature of the array tends to be more uniform. Interestingly, the maximum temperature is not always located at the position of the array center. However, the center and maximum temperature vary with the parameter of the array with a similar law. Here, we propose that the coupling effect of a scattered light field between particles may also be used to regulate temperature distribution. This work provides some theoretical considerations for the field manipulation of nanostructures.

**Supplementary Materials:** The following supporting information can be downloaded at: https://www.mdpi.com/article/10.3390/nano12213924/s1, Figure S1: The absorption and scattering efficiencies of chain composed by 3 Ag nanoparticles with different particle spacing and radius. (a,b) The chain is perpendicular to polarization. (c,d) The chain is parallel to polarization; Figure S2: The absorption and scattering efficiencies of chain composed by 3 Ag nanoparticles with different particle radius and spacing d = 10 nm (a,b) The chain is perpendicular to polarization. (c,d) The chain is parallel to polarization; Figure S3: The absorption and scattering efficiencies of the different structure of Ag nanoparticle with different particle radius (a,b) a single Ag nanoparticle (c,d) 2 × 2 square array of Ag nanoparticles with particle spacing $d$ = 10 nm.

**Author Contributions:** Conceptualization, D.L. and P.H.; methodology, P.H.; software, P.H.; validation, D.L., C.W. and X.L.; investigation, S.W. and P.H.; writing—original draft preparation, P.H.; writing—review and editing, D.L., P.H. and X.L.; supervision, D.L. and C.W.; project administration, S.W. and C.W.; funding acquisition, D.L. All authors have read and agreed to the published version of the manuscript.

**Funding:** This research was funded by Project of Shaanxi Provincial Department of Education (No. 22JC011).

**Data Availability Statement:** Not applicable.

**Acknowledgments:** The authors acknowledge the technical and financial support from Project of Shaanxi Provincial Department of Education.

**Conflicts of Interest:** The authors declare no conflict of interest.

## References

1. Amendola, V.; Pilot, R.; Frasconi, M.; Marago, O.M.; Lati, M.A. Surface plasmon resonance in gold nanoparticles: A review. *J. Phys. Condens. Matter* **2017**, *29*, 203002. [CrossRef]
2. Kelly, K.L.; Coronado, E.; Zhao, L.L.; Schatz, G.C. The optical properties of metal nanoparticles: The influence of size, shape, and dielectric environment. *J. Phys. Chem. B* **2003**, *107*, 668–677. [CrossRef]
3. Sosa, I.O.; Noguez, C.; Barrera, R.G. Optical properties of metal nanoparticles with arbitrary shapes. *J. Phys. Chem. B* **2003**, *107*, 6269–6275. [CrossRef]
4. Ross, M.B.; Mirkin, C.A.; Schatz, G.C. Optical properties of one-, two-, and three-dimensional array of plasmonic nanostructures. *J. Phys. Chem. C* **2016**, *120*, 816–830. [CrossRef]
5. Manjavacas, A.; Zundel, L.; Sanders, S. Analysis of the limits of the near-field produced by nanoparticle arrays. *ACS Nano* **2019**, *13*, 10682–10693. [CrossRef]
6. Luo, D.B.; Shi, B.; Zhu, Q.; Qian, L.L.; Qin, Y.P.; Xie, J.J. Optical properties of Au-Ag nanosphere dimer: Influence of interparticle spacing. *Opt. Commun.* **2020**, *458*, 124746. [CrossRef]
7. Nie, S.M.; Emory, S.R. Probing single molecules and single nanoparticles by surface-enhanced Raman scattering. *Science* **1991**, *275*, 1102–1106. [CrossRef]
8. Okamoto, K.; Niki, I.; Shvartser, A.; Narukawa, Y.; Mukal, T.; Scherer, A. Surface-plasmon-enhanced light emitters based on InGaN quantum wells. *Nat. Mater.* **2004**, *3*, 601–605. [CrossRef]
9. Atwater, H.A.; Polman, A. Plasmonics for improved photovoltaic devices. *Nat. Mater.* **2010**, *9*, 205–213. [CrossRef]
10. Herves, P.; Perez-Lorenzo, M.; Liz-Marzan, L.M.; Dzubiella, J.; Lu, Y.; Ballauff, M. Catalysis by metallic nanoparticles in aqueous solution: Model reactions. *Chem. Soc. Rev.* **2012**, *41*, 5577–5587. [CrossRef]
11. Baffou, G.; Berto, P.; Urena, E.B.; Quidant, R.; Monneret, S.; Polleux, J.; Rigneault, H. Photoinduced heating of nanoparticle arrays. *ACS Nano* **2013**, *7*, 6478–6488. [CrossRef]
12. Baffou, G.; Cichos, F.; Quidant, R. Applications and challenges of thermoplasmonics. *Nat. Mater.* **2020**, *19*, 946–958. [CrossRef]
13. Boyer, D.; Tamarat, P.; Maali, A.; Lounis, B.; Orrit, M. Photothermal imaging of nanometer-sized metal particles among scatterers. *Science* **2002**, *297*, 1160–1163. [CrossRef]
14. Pitsillides, C.M.; Joe, E.K.; Wei, X.B.; Anderson, R.R.; Lin, C.P. Selective cell targeting with light-absorbing microparticles and Nanoparticles. *Biophys. J.* **2003**, *84*, 4023–4032. [CrossRef]
15. Baffou, G. *Thermoplasmonics: Heating Metal Nanoparticles Using Light*; Cambridge University Press: Cambridge, UK, 2017. [CrossRef]
16. Ali, M.R.; Wu, L.; El-Sayed, M.A. Gold nanoparticle-assisted plasmonic photothermal therapy advances towards clinical application. *J. Phys. Chem. C* **2019**, *123*, 15375–15393. [CrossRef]
17. Guerrero, A.R.; Hassan, N.; Escobar, C.A.; Albericio, F.; Kogan, M.J.; Araya, E. Gold nanoparticles for photothermally controlled drug release. *Nanomedicine* **2014**, *9*, 2023–2039. [CrossRef]

18. Chen, M.J.; He, Y.R.; Zhu, J.Q.; Kim, D.R. Enhancement of photo-thermal conversion using gold nanofluids with different particle sizes. *Energy Convers. Manag.* **2016**, *112*, 21–30. [CrossRef]
19. Luo, M.G.; Zhao, J.M.; Liu, L.H.; Antezza, M. Photothermal behavior for two-dimensional nanoparticle ensembles: Multiple scattering and thermal accumulation effects. *Phys. Rev. B* **2022**, *105*, 235431. [CrossRef]
20. Draine, B.T.; Flatau, P.J. Diserete-dipole approximation for periodic targets: Theory and tests. *J. Opt. Soc. Am. A* **2008**, *25*, 2693–2703. [CrossRef]
21. Flatau, P.J.; Draine, B.T. Fast near field calculations in the discrete dipole approximation for regular rectilinear grids. *Opt. Express* **2012**, *20*, 1247–1252. [CrossRef]
22. Siahpoush, V.; Ahmadi-kandjani, S.; Nikniazi, A. Effect of plasmonic coupling on photothermal behavior of random nanoparticles. *Opt. Commun.* **2018**, *420*, 52–58. [CrossRef]
23. Bohren, C.F.; Huffman, D.R. *Absorption and Scattering of Light by Small Particles*; John Wiley & Sons, Inc.: New York, NY, USA, 1998; pp. 69–81. [CrossRef]
24. Johnson, P.B.; Christy, R.W. Optical constants of the noble metals. *Phys. Rev. B* **1972**, *6*, 4370–4379. [CrossRef]
25. Metwally, K.; Mensah, S.; Baffou, G. Isosbestic thermoplasmonic nanostructures. *ACS Photonics* **2017**, *4*, 1544–1551. [CrossRef]
26. Geraci, G.; Hopkins, B.; Miroshnichenko, A.E.; Erkihun, B.; Neshev, D.N.; Kivshar, Y.S.; Maier, S.A.; Rahmani, M. Polarisation-independent enhanced scattering by tailoring asymmetric plasmonic systems. *Nanoscale* **2016**, *8*, 6021–6027. [CrossRef]
27. Baffou, G.; Urena, E.B.; Berto, P.; Monneret, S.; Quidant, R.; Rigneault, H. Deterministic temperature shaping using plasmonic nanoparticle assemblies. *Nanoscale* **2014**, *6*, 8984–8989. [CrossRef]
28. Zhu, M.; Baffou, G.; Meyerbroker, N.; Polleux, J. Micropatterning thermoplasmonic gold nanoarrays to manipulate cell adhesion. *ACS Nano* **2012**, *6*, 7227–7233. [CrossRef]
29. Li, Y.F.; Liu, X.D.; Li, J.; Wu, J. Photothermal forces enhanced by nanoneedle array for nanoparticle manipulation. *Opt. Commun.* **2020**, *454*, 124439. [CrossRef]

 *nanomaterials*

*Review*

# Nonlinear Optical Microscopy and Plasmon Enhancement

Yi Cao [1,2,†], Jing Li [3,†], Mengtao Sun [2,*], Haiyan Liu [1,*] and Lixin Xia [1,*]

1. Liaoning Key Laboratory of Chemical Additive Synthesis and Separation, Yingkou Institute of Technology, Yingkou 115014, China; d202110417@xs.ustb.edu.cn
2. School of Mathematics and Physics, University of Science and Technology Beijing, Beijing 100083, China
3. Key Laboratory of Photochemical Conversion and Optoelectronic Materials, Technology Institution Physical and Chemistry, Chinese Academic Science, Beijing 100190, China; lijingsp@mail.ipc.ac.cn
* Correspondence: mengtaosun@ustb.edu.cn (M.S.); lhy4486@163.com (H.L.); lixinxia@lnu.edu.cn (L.X.)
† These authors contributed equally to this work.

**Abstract:** Improving nonlinear optics efficiency is currently one of the hotspots in modern optical research. Moreover, with the maturity of nonlinear optical microscope systems, more and more biology, materials, medicine, and other related disciplines have higher imaging resolution and detection accuracy requirements for nonlinear optical microscope systems. Surface plasmons of metal nanoparticle structures could confine strong localized electromagnetic fields in their vicinity to generate a new electromagnetic mode, which has been widely used in surface-enhanced Raman scattering, surface-enhanced fluorescence, and photocatalysis. In this review, we summarize the mechanism of nonlinear optical effects and surface plasmons and also review some recent work on plasmon-enhanced nonlinear optical effects. In addition, we present some latest applications of nonlinear optical microscopy system research.

**Keywords:** plasmon; nonlinear optics; microscopy systems

## 1. Introduction

As one of the essential branches of modern physical photonics, the discovery of nonlinear optics benefits from the invention of lasers [1,2]. Nonlinear optics takes the physical phenomena and reality applications resulting from the interaction between glaring light and nonlinear materials as research goals, including the generation of harmonics, optical detection, frequency modulation, etc. [3]. In 1961, Fraken produced a light with a wavelength of 347 nm by irradiating a crystal with a ruby laser [4]. The wavelength of this new light is not the same as the wavelength of the incident light and is exactly half the wavelength of the incident light. This finding from the experiment verifies the second harmonic generation (SHG) for the first time. Since then, Schawlow et al., have also theoretically analyzed nonlinear optical phenomena such as quasi-phase matching and phase matching in the resonant cavity [5,6]. Effects such as stimulated light scattering, two-photon absorption, and optical Kerr effect are found in early nonlinear optics research [7–9]. The early discovery and exploration of nonlinear optical phenomena by researchers laid the foundation for the development of the field [10]. Based on the advantages of nonlinear optical effects, photonic devices such as nonlinear optical microscopes have matured [11,12]. However, due to the requirement of the high-intensity electric field, the inherent nonlinear optical response of the material is not obvious. Some nonlinear optical crystals usually use complementary ways to increase conversion efficiency, but they are challenging to use in fabricating integrated optoelectronic devices due to size constraints [13]. Therefore, how to improve the nonlinear response and conversion efficiency in the nanoscale is the current barrier to nonlinear optics research.

In 1921, Albert Einstein proved and revealed the existence of the photoelectric effect through experiments and won the Nobel Prize in Physics, which opened the research

chapter of modern photoelectric interaction [14]. The unique reflective luster of the noble metal surface is closely related to the metal's unique optical properties and the free charge shifting inside the metal [15]. When electromagnetic waves of a specific wavelength irradiate the metal nanostructures, the plasmon effect generated on the surface can bind the electromagnetic waves to the vicinity of the metal nanostructures [16]. Plasmonic nanostructures could control linear and nonlinear optical processes and enhance the interaction of light and material by confining electromagnetic waves in the subwavelength range, which has a wide range of application prospects in the fields of single-molecule detection [17], super-resolution detection [18], light capture and emission [19], and optical force manipulation [20]. Meanwhile, the plasmon effect's modulation of nonlinear optical processes can also help improve the nonlinear optical effect under low-light conditions and reduce the size of nonlinear optical devices [21]. The discovery of these traits has a tremendous guiding effect on the research and development of integrated optoelectronic devices and has aroused the interest of researchers [22]. The generation of nonlinear optical effects relies on high-intensity external electromagnetic fields. Surface plasmons can trap light in nanostructures in free space and form huge local electromagnetic field enhancements [23]. Combining surface plasmons with nonlinear optical effects can realize nonlinear effects under low-light conditions and has excellent potential for developing nonlinear sensing and all-optical regulation.

This review starts from nonlinear optics and surface plasmons' characteristics and introduces the research progress on the nonlinear optical effects of surface plasmon metal nanostructures. In addition, we also review the experimental results of the combined use of nonlinear optical microscopy systems and prospect the emerging application prospects of this field and the potential research directions for the existence of nonlinear optics in nanostructures.

## 2. Nonlinear Optics Effect Process of Surface Plasmon

Under the excitation of an external electric field, the free electrons located in the conduction band around the Fermi level in the interior of the metal nanoparticle will collectively oscillate on the particle surface. The electric field energy in the charge resonance state will be converted into the kinetic energy of the collective oscillation of free electrons on the particle surface and generate the surface plasmon (SP) effect, which includes the surface plasmon polariton (SPP) and localized surface plasmon (LSP) resonance [24]. Localized surface plasmon resonance refers to the collective oscillation of surface free electrons for metal nanoparticles under the excitation of an external field, resulting in boundary conditions due to the size and shape of the nanoparticles. Free electrons could resonate and enhance the local electric field strength around the metal nanoparticles under specific excitation conditions. The LSP peak intensity and position are affected by the morphology, size, material type, and dielectric constant of the metal nanoparticles, resulting in displacement or intensity change [25]. Assuming that there is an isotropic metal sphere with radius d inside the isotropic medium, and the radius $d$ of the sphere is much smaller than the incident wavelength $\lambda$, it can be approximated that the electric field around the metal sphere is the same everywhere. In the other words, the model is a quasi-static approximation [26]. The extinction cross-section ($\sigma_{ext}$) of the metal sphere is shown in Equation (1):

$$\sigma_{ext} = \sigma_{abs} + \sigma_{sca} = k\pi d^3 12\varepsilon_d^{3/2} \left( \frac{\varepsilon_m''}{(\varepsilon_m' + 2\varepsilon_d)^2 + \varepsilon_m''^2} \right) \quad (1)$$

where the wave vector $k = 2\pi/\lambda$. $\sigma_{sca}$ and $\sigma_{abs}$ are the scattering cross-section and absorption cross-section of the metal sphere, respectively. $\varepsilon_m$ and $\varepsilon_d$, respectively, represent the complex permittivity of the metal sphere and the medium. When the $\sigma_{ext}$ reach the most significant situation, most of the incident light is absorbed or scattered, which is the LSP resonance state. Complex permittivity of the metal sphere $\varepsilon_m = \varepsilon(\omega, d) = \varepsilon_m'(\omega, d) + i\varepsilon_m''(\omega, d)$, which means the complex permittivity of the metal sphere is closely related to the frequency of incident light and the size of metal particles. In addition, the imaginary and real parts of the complex permittivity correspond to the position and width of the LSP resonance peak, respectively. This indicates that the LSP resonance peak position and full half-high width could be regulated by changing the complex permittivity of the metal particles, which provides a way to study the controllability of LSP resonance. Researchers usually use chemical synthesis to prepare metal nanoparticles with controllable size and morphology and find that the curvature of metal nanoparticles is closely related to the enhancement of the local electric field [27]. In particular, nanotips, protrusions, and other "antenna"-like shapes have strong electromagnetic hot spots around them. In recent years, researchers have also found that the coupling effect of metallic nanogap can cause an extreme electric field enhancement in this region, which as shown in Figure 1. The local field strength variation law of electromagnetic coupling "hot spots" between metal nanoparticles by controlling the distance between metal nanoparticles [28].

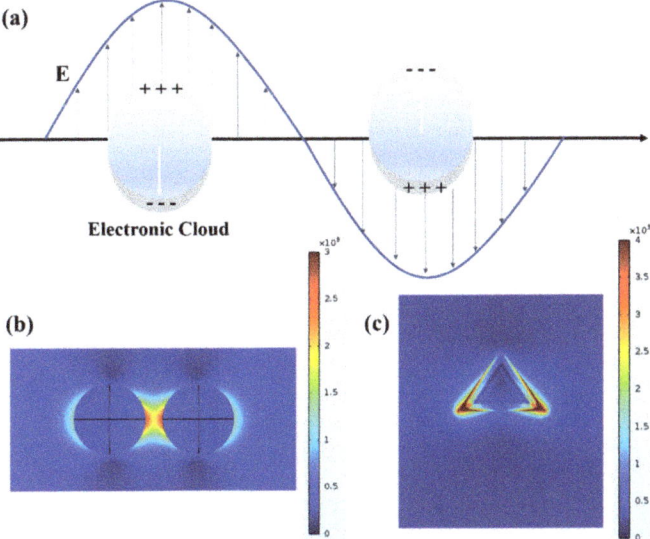

**Figure 1.** (a) Localized surface plasmon resonance on the surface of metal nanoparticles. (b) The hot spot with nanogap is when the two Au nanospheres with a diameter of 20 nm are separated by 3 nm. (c) The "tip-hot spot" effect at the tip of Au nanocones with regular tetrahedral structure.

Unlike LSP, surface plasmon polaritons refer to the resonant coupling of free electrons and incident photons on the surface of metal substrates with periodic structures such as metal nanodiscs and gratings to form an electromagnetic mode. In this case, evanescent waves are generated at the dielectric interface and propagate parallel to the interface [29]. SPP refers to a new hybrid electromagnetic mode generated by the resonant coupling of incident photons of an external field and free electrons on the surface of metal nanosubstrates, which propagates horizontally along with the interface between the surface of the metal substrate and the mediu [30]. The electric field amplitude decays exponentially along the vertical direction of the interface. In general, the field attenuation of air or glass above the metal is 1/2 of the incident light wavelength, and the metal's attenuation depends on the surface depth of the metal. The real ($K_r$) and imaginary ($K_i$) parts of the SPP wave vector can be represented by Equations (2) and (3):

$$K_r = \frac{\omega}{c}\left(\frac{\varepsilon'_m \varepsilon_d}{\varepsilon'_m + \varepsilon_d}\right)^{1/2} \qquad (2)$$

$$K_i = \frac{\omega}{c} \frac{\varepsilon''_m}{2{\varepsilon'_m}^2} \left( \frac{\varepsilon'_m \varepsilon_d}{\varepsilon'_m + \varepsilon_d} \right)^{3/2} \quad (3)$$

where $\omega$ is the angular frequency of the incident light and $c$ is the speed of light. $\varepsilon'_m$ and $\varepsilon''_m$ are the real and imaginary parts of the metal dielectric constant and $\varepsilon_d$ is the dielectric constant of the medium. If the frequencies are the same, the plasmon wave vector on the metal surface is larger than the light wave vector ($K_{SPP} > \frac{\omega}{c}$). There will generate the SPP effect at the interface between metal and medium. Therefore, we must use the wave vector matching method to realize the resonance coupling excitation of surface plasmon and incident light.

The incident light wave must be excited on the metal surface to generate surface plasmon polaritons so that the effective control of the SPP mode can be accomplished. According to the SPP dispersion relationship in Figure 2c, it could be shown that at the same frequency, the photon wave vector is smaller than the surface plasmon polariton wave vector generated by exciting the metal surface [31]. Taken as a whole, only by increasing the wave vector of the incident light wave can the surface plasmon polaritons be excited on the metal surface.

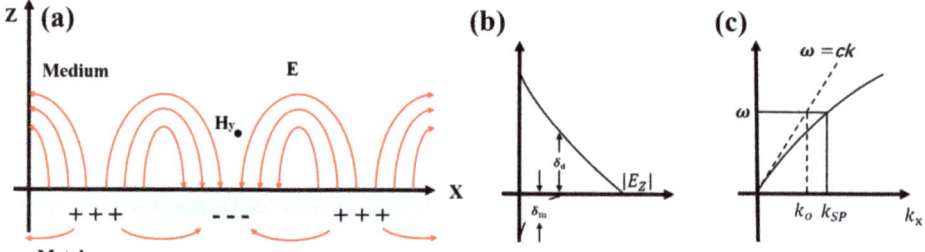

**Figure 2.** Electric field properties of SPP at the interface between dielectric and metal surface: (**a**) Coupling and electronic oscillation of SPP electromagnetic waves. (**b**) The electric field component of the SPP is bound on both sides of the interface and decays exponentially. (**c**) Dispersion plot of SPP.

The necessary condition for the generation of nonlinear optical effects is the interaction of high-intensity light with optic materials [32]. Nonlinearity in nonlinear optics means that the response of a substance to an external light field is in the form of a nonlinear equation, and is closely related to the intensity of the external light field [33]. For a material system, the relationship between the external light field intensity $E(t)$ and the polarization intensity $P(t)$ is shown by Equation (4):

$$P(t) = P^{(1)}(t) + P^{(2)}(t) + P^{(3)}(t) + \cdots = \varepsilon \left[ \chi^{(1)} E^1(t) + \chi^{(2)} E^2(t) + \chi^{(3)} E^3(t) + \cdots \right] \quad (4)$$

where $\varepsilon$ is the dielectric constant in vacuum, and $\chi^{(n)}$ is the nonlinear polarizability of the $n$ order. $P^{(1)}(t)$ is the linear part, where the resonance amplitude of the applied electric field is linearly related to the optical response of the materials. As the intensity of the applied light field increases, the nonlinear response of $P(t)$ to the light field begins to appear gradually. Among them, the second-order nonlinear optical effect is closely related to $\chi^{(2)}$, which includes the second harmonic generation (SHG), sum-frequency generation (SFG), and difference frequency generation (DFG). The third-order nonlinear optical effects originating from polarizability $\chi^{(3)}$ mainly include triple harmonics generation (THG), four-wave mixing (FWM), coherent anti-Stokes Raman scattering (CASR), and so on. $\chi^{(n)}$ ($5 \leq n$) is the higher-order nonlinear optical process that usually produces higher-order harmonics generation (HHG). Among the common nonlinear optical processes, the nonlinear processes of $\chi^{(2)}$ and $\chi^{(3)}$ are relatively common, and the localized electromagnetic field enhancement caused by surface plasmon resonance could improve the conversion efficiency in nonlinear optics, so that has received extensive attention from researchers.

## 3. Surface Plasmon Resonances Enhanced the Nonlinear Optics Microscopy
### 3.1. Nonlinear Optical Enhancement in Surface Plasmons of Metal Nanoparticles

The essence of using surface plasmon resonance to enhance nonlinear optical effects is to match the frequency of the excitation light or frequency-doubling light with the surface plasmon resonance frequency to achieve the effect of excitation or emission enhancement [34]. Based on this condition, a variety of metal nanostructures were prepared by researchers, such as noble metal dimer structure [35,36], metal nanograting structure [37], metal nanocube structure [38], vertical Au nanorods (AuNRs) array structure, and so on [39,40]. Furthermore, for even-order nonlinear optical effects (e.g., SHG), the symmetrical structure of the material is essential [41]. Mi et al. [42] found that multi-surface plasmon resonance (MSPR) induced by Au@Ag NRs structure with frequency doubling effect could enhance CARS and two-photon excited fluorescence (TPEF) in nonlinear optical microscopy. By controlling the ratio of Hexadecyltrimethyl-ammonium bromide (CTAB), $HAuCl_4$, and $AgNO_3$, they prepared Au@Ag NRs with different aspect ratios by wet chemical reduction method under a specific temperature and humidity environment. They successfully realized the frequency-doubling relationship of the UV-vis absorption peak positions of Au@Ag NRs. The fundamental and doubling peaks of Au@Ag NRs are at 800 nm and 400 nm, respectively. Meanwhile, they also demonstrated the enhancement effect of MSPR on nonlinear optical processes by TPEF and two-photon CARS characterization images of the two-dimensional (2D) material $g-C_3N_4$.

As shown in Figure 3a–c, they successfully coated a 10 nm-thick Ag shell on the surface of AuNR to form Au@Ag NRs. The length of the inner AuNR is 100 nm, and the diameter of the AuNR is 12 nm. Different from AuNRs, the emerging core-shell NR structure constructed by Au@Ag NRs will blueshift the resonance peak of AuNRs from 910 nm and 520 nm to 800 nm and 500 nm, and generate a new resonance peak at 400 nm, which is shown in Figure 3d. As a typical nonlinear optical phenomenon, TPEF approximates two high-intensity lower-frequency photons as one higher-frequency photon (frequency doubling relationship) to excite the fluorescence signal of the materials. The absorption peak of $g-C_3N_4$ in the 2D material is at ~400 nm, and the fluorescence emission peak is at ~450 nm, as shown in Figure 4b. Figure 4a is an optical microscopy image of $g-C_3N_4$ monolayer. Figure 4c,d are two-photon excitation (800 nm) fluorescence microscopy images of $g-C_3N_4$ and Au@Ag NRs/$g-C_3N_4$. Comparing the microscopic imaging results, it could be found that in Figure 4c, $g-C_3N_4$ only produces fluorescence under local focus, and the intensity is weak. In Figure 4d, there are Au@Ag NRs particles on the surface of $g-C_3N_4$. Based on the enhanced local electric field generated by the surface plasmon of Au@Ag NRs, the fluorescence intensity of $g-C_3N_4$ was significantly enhanced. The frequency-doubling absorption peak of Au@Ag NRs just matches that of $g-C_3N_4$, and the two photons could be absorbed in this band that realizes the fluorescence enhancement of $g-C_3N_4$. In addition, they also investigated the enhancement of CARS characterization of $g-C_3N_4$ by Au@Ag NRs, which is another nonlinear optical effect. The CARS and TPEF nonlinear optical signals of $g-C_3N_4$ can be distinguished after plasmon enhancement of Au@Ag NRs, because the frequency doubling of Au@Ag NRs at 800 and 400 nm enhances the non-linear optical signals. The simulations in COMSOL software strongly support the experimental results. At an incident angle of 30°, the enhancement factors (EF) of Au@Ag NRs surface plasmons for CARS and TPEF can reach $1.6 \times 10^4$ and $6 \times 10^{16}$, respectively, which is huge for the nonlinear optical signal enhancement of $g-C_3N_4$ alone. This research not only enables the enhancement of nonlinear optic signals by frequency doubling of metal nanorods but also demonstrates the great potential of MSPR for accurate characterization and improved imaging resolution in the future.

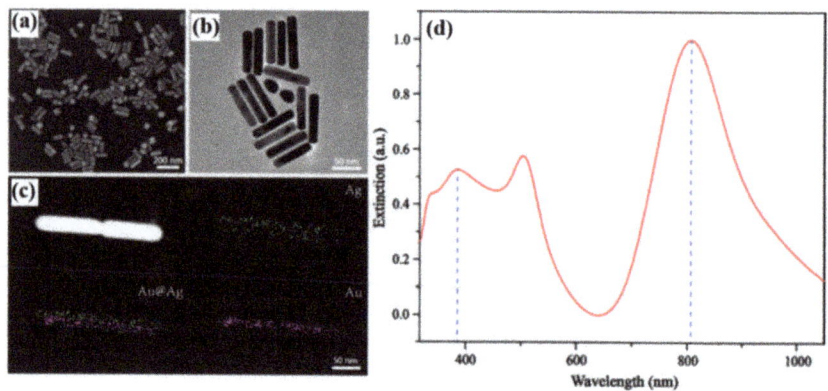

**Figure 3.** (a,b) Characterization image of Au@Ag NRs, respectively, by the scanning electron microscope (SEM) and transmission electron microscope (TEM). (c) Characterization results of different elemental compositions of Au@Ag NRs. (d) UV-Vis absorption spectra of experimentally synthesized Au@Ag NRs [42]. Republished with permission from Ref. [42]. Copyright 2022 Walter de Gruyter.

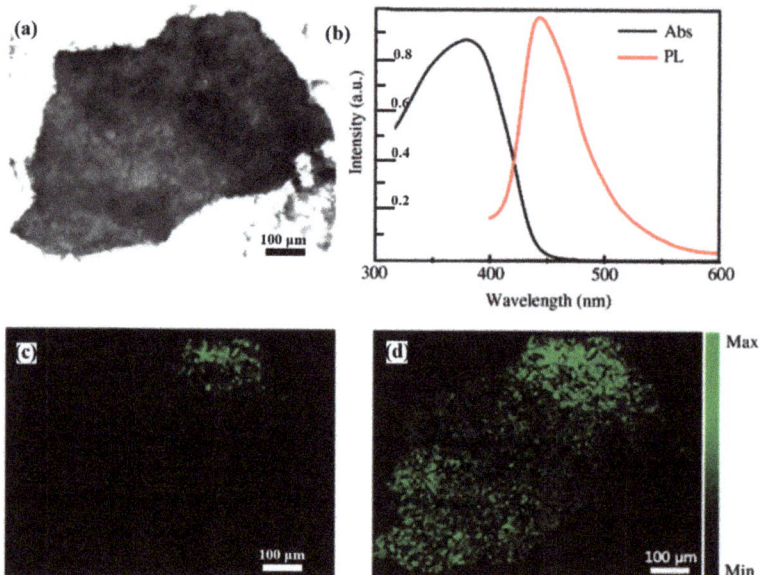

**Figure 4.** (a) Optical brightfield image of g-$C_3N_4$ without Au@Ag NRs. (b) Absorption and fluorescence (PL) spectra of g-$C_3N_4$. (c,d) are the TPEF images of g-$C_3N_4$ without and with Au@Ag NRs on the surface, respectively [42]. Republished with permission from Ref. [42]. Copyright 2022 Walter de Gruyter.

Based on the frequency-doubling relationship of Au@Ag NRs, Cui et al. successfully realized the nonlinear microscopic characterization of CARS and TPEF of plasmon-enhanced non-fluorescent microorganisms according to previous research, which is shown in Figure 5 [43]. They found that Au@Ag NRs with frequency-doubled surface plasmon resonance (SPR) peaks (400 nm/800 nm) could effectively induce the linear fluorescence signal of *E. coli* and also enhance the nonlinear spectral signal of its TPEF. Besides, the CARS nonlinear spectral signals of *E. coli* and *S. aureus* can also be effectively enhanced by the SPR effect of Au@Ag NRs. Figure 5a,b are the TEM characterization results of the

Au@Ag NRs parallel array after the self-assembly experiment and the *E. coli* incubated in Au@Ag NRs solution, respectively. Between each self-assembly, Au@Ag NRs parallel array nanogap would generate the hotspot, which would enhance the local intensity of the electromagnetic field around the nanorod. They first performed confocal imaging of two standard CARS peaks of *E. coli* at 1094 cm$^{-1}$ and 2993 cm$^{-1}$ with light at 532 nm, as shown in Figure 5c,d. The left side of the blue dotted line is the colony area without Au@Ag NRs, and the right side is the colony area with Au@Ag NRs.

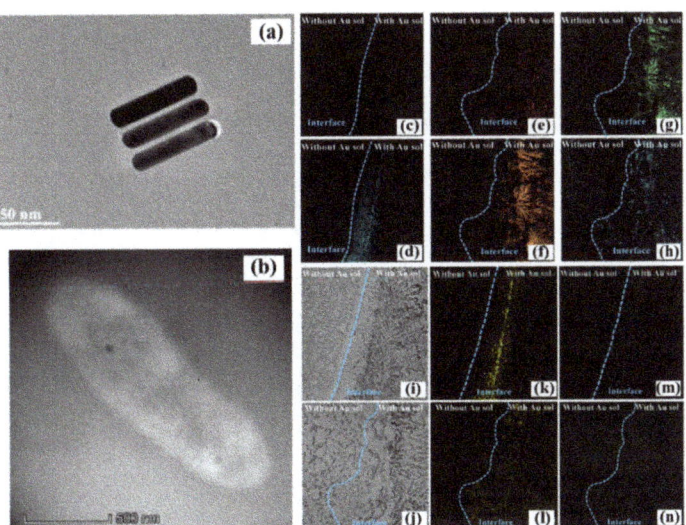

**Figure 5.** (**a**,**b**) are the TEM images of the self-assembled Au@Ag NRs and *E. coli*, respectively. (**c**,**d**) are the results of CARS imaging of *E. coli* at 1094 cm$^{-1}$ and 2933 cm$^{-1}$, respectively. (**e**–**h**) are the CARS imaging results of *S. aureus* at 1094 cm$^{-1}$ (**e**), 1340 cm$^{-1}$ (**f**), 1589 cm$^{-1}$ (**g**) and 2933 cm$^{-1}$ (**h**), respectively. (**i**–**n**) are the TPEF imaging signals of *E. coli* and *S. aureus* under brightfield microscopy (**i**,**j**), 570–630 nm (**k**,**l**), and 495–540 nm (**m**,**n**) channel regions, respectively. In the nonlinear microscopic imaging image, the right and left sides of the blue dotted line are the regions with and without Au@Ag NRs, respectively [43]. Republished with permission from Ref. [43]. Copyright 2022 Elsevier.

The imaging results clearly show that the nonlinear signal of *E. coli* CARS enhanced by the metal surface plasmon is more obvious, and the imaging resolution of the colony is greatly improved. They then also performed confocal imaging of the characteristic CARS signal of *S. aureus*. Similar results are shown in Figure 5e–h, the CARS imaging intensity on one side (right side of the blue dotted line) coated with Au@Ag NRs is significantly higher than that on the other side. This is because the SPR intensity of Au@Ag NRs is strongest at 800 nm, which can well enhance the different light in CARS. Meanwhile, it also proves that the surface plasmon of Au@Ag NRs indeed enhances the CARS signal. In order to better prove that plasmons can enhance nonlinear optical efficiency, they also verified the TPEF process of *E. coli* and *S. aureus* by using the plasmon frequency doubling effect of Au@Ag NRs. Analyzing the microbial environment and activity using CARS technology is cumbersome, and the fluorescent metabolite signals of microorganisms are more commonly used in research to confirm the life state of microorganisms. Figure 5i,j, respectively, are bright-field microscopic images of *E. coli* and *S. aureus*, from which the distribution of colonies and the effect of Au@Ag NRs on colony growth could be observed. Figure 5k–n are the TPEF confocal images of *E. Coli* and *S. Aureus* in the 575 nm to 630 nm and 495 nm to 540 nm channels under two-photon excitation at 800 nm, respectively. The blue dotted line is also used as the coverage with or without Au@Ag NRs. The imaging results could clearly demonstrate the signal amplification in the nonlinear optical process of two-photon fluorescence imaging by plasmon-enhanced. The research results provide a promising prospect for applying frequency-doubling metal nanoparticles in plasmon-enhanced nonlinear optical processes and microscopic imaging.

Shen et al. [44] designed a reusable plasmon-enhanced SHG (PESHG) substrate suitable for near-ultraviolet (NUV) wavelengths synthesizing Ag mushroom arrays. As shown in Figure 6a, they fabricated Ag mushroom arrays on Si/Au film substrates using nanolithography imprinting technology and the electrochemical deposition process. Figure 6b,c are the SEM characterization results of this array, which intuitively illustrate the uniformity of the array, and this periodic structure is the key to the SPP effect. In addition, controlling the nanogap spacing between mushrooms also helped generate more electromagnetic hotspots. Metal mushroom arrays are characterized by reducing their centrosymmetry at the three-dimensional level and can significantly enhance the SHG signal. Compared with the SHG signal generated by the surface of Au film alone, the SHG signal released by the Au mushroom array (GMA) was enhanced by 13 times, as shown in Figure 6d. Benefiting from the superior metal activity of Ag compared to Au, they deposited an Ag film on the surface of the Au mushroom array (SMA). They found that the new Ag mushroom array improved the SHG signal significantly. Under different fundamental frequencies, SMA has improved SHG signal than GMA, as shown in Figure 6e, especially at the fundamental wavelength of 860 nm, it has a nearly 80-fold improvement. This finding has guiding significance for applying SHG in nonlinear optics, such as optical communication and surface detection.

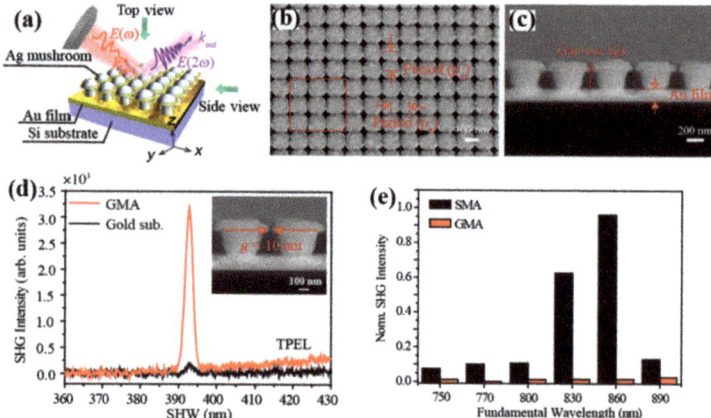

**Figure 6.** (a) Schematic diagram of the fabrication of metal mushroom arrays. (b) The top view of SEM characterization of SMA. (c) A single Ag mushroom has a height of 220 nm and a diameter of 300 nm, which is the side view. (d) Signal comparison of GMA and SHG on the surface of Au film substrate. TPEL is expressed as two-photon excitation should. Inset is the SEM characterization of GMA. (e) Fundamental wavelength-dependent comparison of PESHG performance of SMA and GMA [44]. Republished with permission from Ref. [44]. Copyright 2022 American Physical Society.

Apart from the surface plasmon effect produced by common metal nanoparticles (nanorods, nanoparticles . . . ), metal tips as a typical metal structure, produce a huge localized electromagnetic field focus effect, which is widely used for scanning probe microscopy (SPM), tip-enhanced Raman scattering (TERS) and other research [45,46]. Jiang et al. [47] achieved coherent nonlinear imaging and graphene nanospectroscopy detection by means of metal tip excitation. They achieved the SPP effect by etching the periodic grating structure on the outer ring of the Au tip, as shown in Figure 7a, and a huge local electromagnetic field was generated at the Au tip with a radius of 10 nm, so they were able to detect the four-wave mixing (FWM) response of graphene. Figure 7b shows the imaging results of different layers $n$ of graphene according to different response strengths.

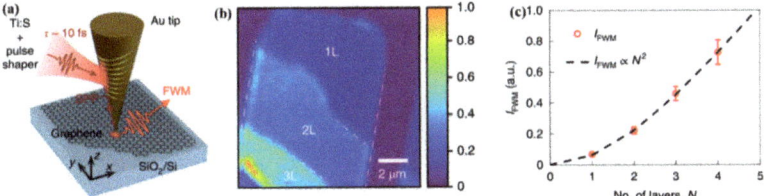

**Figure 7.** (a) Schematic illustration of the Au tip after grating etching for localized FWM excitation. (b) Near-field FWM imaging of monolayer and bi- and tri-layer graphene. The white dotted line is the edge of the graphene. (c) Fitted model of the dependence of thickness ($n$) on the intensity of integrated FWM ($I_{FWM}$) [47]. Republished with permission from Ref. [47]. Copyright 2022 Springer Nature.

As $n$ increases, the strength ($I_{FWM}$) of the FWM then exhibits a quadratic dependence ($I_{FWM} \propto N^2$), which is shown in Figure 7c. Based on this research, the number of graphene layers could be judged according to the value of $I_{FWM}$. More essentially, this study demonstrates the feasibility of near-field nonlinear optics research in the nanometer range. Among them, the increase in nonlinear signals due to nano-focusing and near-field enhancement enables the further development of nonlinear nano-optics of two-dimensional materials and their nanostructures. This also makes it possible to finally develop nonlinear optical integrated devices with ultra-high sensitivity.

$MoS_2$ as a transition metal dichalcogenide (TMDs), has similar physical structure and chemical properties to the typical two-dimensional material graphene and is currently a research hotspot in the field of semiconductors and two-dimensional materials. Wang et al. used a finite-difference time-domain (FDTD) method to investigate the properties of the $MoS_2$ monolayer with the Ag tip to enhance two-photon-excited fluorescence (2 pF) [48]. Figure 8a shows the model structure of $MoS_2$ tip-enhanced by 2 pF. The incident light in the model is irradiated to the $MoS_2$ surface with a thickness of 1 nm on the Ag substrate by oblique incidence. The inset is a detailed schematic of the model. Among them, the incident angle is θ, the tip radius is r, the tip full cone angle is β, and the tip inclination angle and spacing are denoted by α and d, respectively. In the general fluorescence process, due to the low photon density of the incident light, a fluorescent molecule can only absorb one photon simultaneously and then emit a fluorescent photon through the radiation transition, which is called single-photon fluorescence.

For the two-photon fluorescence process, the intensity of the excitation light source is relatively high, and the photon density meets the requirement of simultaneous absorption of two photons by fluorescent molecules. In the normal two-photon fluorescence emission process, the photon density is insufficient to produce two-photon absorption, so a femtosecond pulsed laser is usually used, and its instantaneous power can reach the order of megawatts. The wavelength of two-photon fluorescence is shorter than that of the excitation light, but the photon energy density is higher, which is equivalent to the effect produced by half-excitation wavelength excitation, which is shown in Figure 8b, which is a schematic diagram of the energy level of 2 pF. Compared with the single-photon fluorescence emission, the high-order nonlinear optical process in the tip-enhanced 2 pF process leads to the greatly enhanced fluorescence emission. As shown in Figure 8c,d, the 2 pF overall enhancement factor of $MoS_2$ under Ag tip enhancement is much higher than 1 pF, and a distinct fluorescence emission peak can be observed at the 630 nm exciton transition. To enable resonance matching of the SPR with the exciton transition process at 677 nm, a maximum enhancement of 2 pF at r = 65 nm was achieved by varying the radius of the Ag tip (25 nm to 65 nm) and was 41 for $EF_{2\,pF}/EF_{1\,pF}$. This shows that the strength of 1 pF can be increased by at least 40 times using the 2 pF process of the tip to $MoS_2$. These works also show that the metal tip in the SPM can improve the conversion efficiency of nonlinear optical processes, achieve coupling, and enhance the strength of nonlinear optical signals under specific conditions.

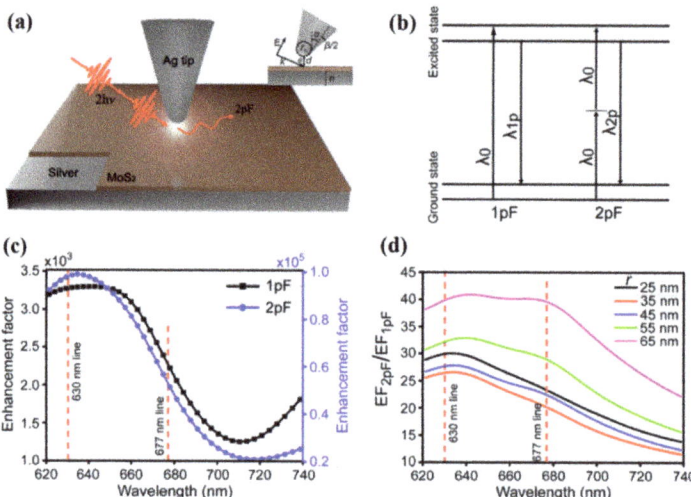

**Figure 8.** (a) Model of the 2 pF process on the Ag tip with the $MoS_2$ surface. (b) Energy level diagrams of excitation and emission processes of 1 pF and 2 pF. (c) Comparison in enhancement factors of 1 pF and 2 pF on $MoS_2$ surface. (d) The change in the Ag tip radius couples the SPR and enhances the exciton emission peak at 677 nm [48]. Republished with permission from Ref. [48]. Copyright 2022 Elsevier.

### 3.2. The Application of Nonlinear Optical Microscopy System

Nonlinear optical microscopy, as a type of laser scanning microscopy, is to image the optical signals generated by the nonlinear interaction between incident light and the sample. There is a nonlinear relationship between the size of the imaged optical signal and the intensity of the incident light. Compared with traditional linear optical microscopes, nonlinear optical microscopes have the advantages of long-wavelength excitation, high spatial resolution, and ultrafast lasers as light sources that could reduce the average power of laser pulses. In recent years, nonlinear optical microscopy systems (such as CARS, SHG, and TPEF microscopy imaging techniques) have been widely used in biological sciences, medical research, and nanomaterials research and development [49]. Among them, the study of SHG signals on the surface of transition metal dichalcogenides (TMDCs) with a certain torsion angle and at the heterojunction based on the polarization-dependent nonlinear microscopies presents a possibility for the application of nonlinear optical microscopy in surface optical inspection [50,51]. In this section, we focus on reviewing the work of nonlinear optical systems in practical applications.

Li et al. imaged porous carbon structures by building the nonlinear optical microscopy system [52]. The specific optical path of the nonlinear optical microscope system is shown in Figure 9b. Figure 9a shows an optical image of the porous carbon after the removal of the nickel structure, in which the synthesis of the porous carbon is completed on the surface of the nickel structure, and the nickel structure has been removed in the optical measurement. The inset in Figure 9a is the Raman signal measured at five different points in the optical image, which can demonstrate the reproducibility and homogeneity of the Raman spectrum of the porous carbon material. Figure 9c,d are the bright-field images and SHG imaging results of the porous carbon material, respectively. The structure of the porous carbon material can be clearly distinguished by comparison. Optical transitions and scattering can be performed in non-centrosymmetric media due to nonlinear optical effects. The stacking of multilayer porous carbon structures could induce SHG generation. Moreover, when the multilayer material is grown epitaxially, the perturbation of the Dirac cone causes the bandgap to open. This indicates that the centrosymmetry is broken, and thus the second-order nonlinear optical response of the porous carbon material can be enhanced. Figure 9e,f, respectively, are the CARS imaging results at 1587 cm$^{-1}$ and 1360 cm$^{-1}$. Figure 9g shows the imaging signal and image structure of the porous carbon material TPEF, which is clearly showing the porous carbon material structure. The transfer of photon momentum in TPEF to the electron system is also demonstrated. Figure 9h is the combined image of Figure 9a–e. The upper and lower figures in Figure 9i are the characterization comparison of CARS and Raman spectra of porous carbon materials, respectively. The nonlinear optical microscope system can reveal that the optical properties of porous carbon materials can be well revealed by the nonlinear optical characterization results of CARS, SHG, and TPEF by the nonlinear optical microscope system. The nonlinear optical microscopy system also provides a potential channel for studying other micro and nanoscale materials. Sun et al. studied the nonlinear optical response of single-layer graphene based on this nonlinear optical system [53]. They found that these nonlinear microscopes could not only clearly observe the morphology and structure of single-layer graphene but also effectively evaluate the quality of graphene. This research shows unique advantages in applying 2D material characterization and biological and medical research.

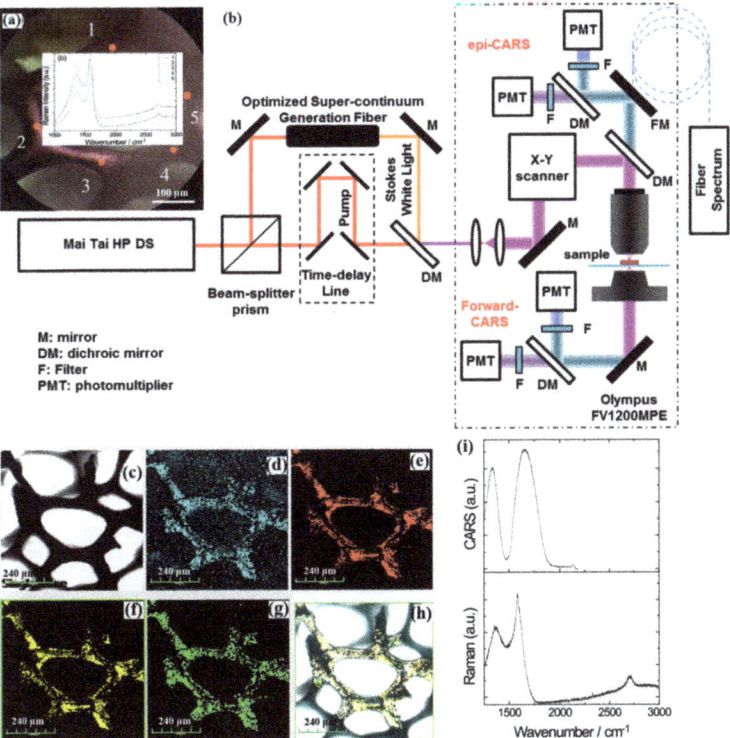

**Figure 9.** (a) Bright-field microscopic image of porous carbon after thin nickel structure. Insets are Raman spectral images at five different points. (b) Optical path diagram of the nonlinear optical microscope system. (c) Brightfield images of porous carbon materials. (d) SHG image of porous carbon materials. (e,f) are CARS images at 1587 cm$^{-1}$ and 1360 cm$^{-1}$, respectively. (g) is a TPEF image of the porous carbon material. (h) is the merged image of Figure 9a–e. (i) compares porous carbon material CARS with Raman spectroscopy [52]. Republished with permission from Ref. [52]. Copyright 2022 Elsevier.

In the process of medical diagnosis and treatment, drug screening and avoidance of side effects are essential research directions. However, traditional screening processes require molecular labeling and detection of fluorescence data. This method of detection carries the risk of altered biological activity. Francesco et al. developed a label-free imaging system based on nonlinear optical microscopy [54]. The nonlinear imaging microscope system combines CARS and Bessel beams to ensure high-resolution spectral imaging and chemical composition analysis at the same time. Based on the nonlinear optical microscopy system, they studied the side effects of drug-induced lipid storage in liver tissue, and successfully extracted the spectral and spatial distribution of lipid and protein components, providing a new channel for detecting drug reaction side effects. Li et al. used nonlinear optical microscopy to observe in situ the symbiosis and competition between collagen and bone during biological evolution [55]. Using nonlinear optical microscopy, they successfully observed that $sdsCO_3^{-2}$ and $PO_3^{-2}$ induced collagen production and bone formation. As bones mature, collagen is gradually broken down from around the bones. Using CARS and SHG nonlinear optical microscopy techniques, the evolution and degradation of bone and collagen can be monitored during snail growth. Figure 10 shows the results of in situ characterization of snail bone and collagen from the first, three, five, and nine days after birth. As shown in Figure 10a,b, a large number of $sdsCO_3^{-2}$ and $PO_3^{-2}$ on day 1 are clustering together with the CARS imaging, while the SHG signal of collagen is not very obvious. The inset is the brightfield microscopic image of the snail corresponding to the day. In the CARS imaging results of $sdsCO_3^{-2}$ and $PO_3^{-2}$ on day 3, it can be found that the aggregated $sdsCO_3^{-2}$ and $PO_3^{-2}$ are forming bone contours. At the same time, collagen also grows along the outline of the bone with the aggregation of $sdsCO_3^{-2}$ and $PO_3^{-2}$. On day 5, the outline of the snail's bones gradually became clear, and the texture began to become apparent. With the rapid development of bones currently, the content of collagen also reaches its peak. On day 9, the snail's skeletal outlines were clearly visible in the CARS imaging, and the circular outlines were fully visible. However, due to the inverse feedback and biological competition mechanism, the SHG signal of collagen at this time degenerates and disappears due to bone maturation. The experimental results brilliantly reveal that the biological evolution of snail bones and collagen has been imaged over time. Based on the study of the synergistic and competitive relationship between collagen and bone growth, it is helpful to explore the repair mechanism of fractures or other bone diseases. The synergistic imaging of these nonlinear optical effects in vivo is helpful to observe the interaction between various substances in the organism.

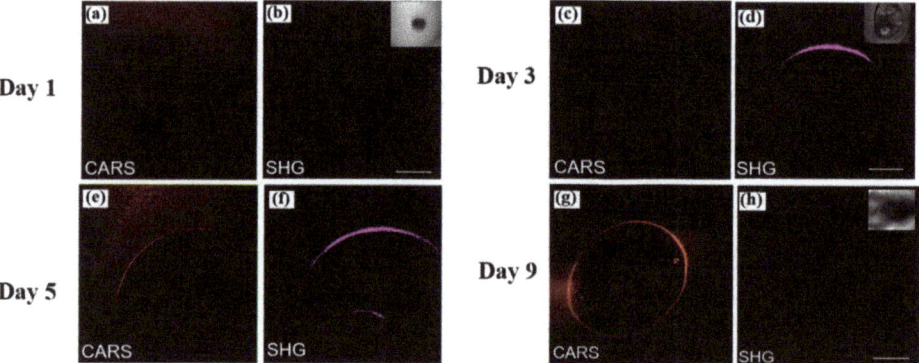

**Figure 10.** (a,b), (c,d), (e,f), (g,h) are the imaging results of CARS of snail bone elements and SHG of collagen on days 1, 3, 5, 9, respectively. Inset shows brightfield microscopy images of snail evolution on the corresponding day [55]. Republished with permission from Ref. [55]. Copyright 2022 John Wiley and Sons.

In order to improve the imaging and detection level, Andrew et al. proposed a CARS nonlinear optical system based on Mid-IR assisted (MIRA) [56]. The nonlinear optical system can effectively obtain coherent Raman scattering signals in a low-noise background, which making the microscope system a potential option for highly sensitive spectral characterization and high-precision imaging. Figure 11a is a schematic diagram of the optical path of the entire nonlinear optical microscope system. They focused two optical signals (532 nm and IR) into methane gas, and then coupled the Raman signal generated after excitation into a spectrometer for analysis. Taking the amplitude of $\chi^{(3)}$ as the detuning function of IR-enhanced Raman and CARS can reflect the enhancement of Raman signal more realistically. The result after they added $\chi^{(3)}$ to the Raman signal is shown in Figure 11b. Compared with traditional CARS, MIRA-based CARS can more effectively couple the ground and excited states. With the advantages of this system, the MIRA-based CARS nonlinear optical microscope system provides a new solution for spectral detection and imaging with high sensitivity.

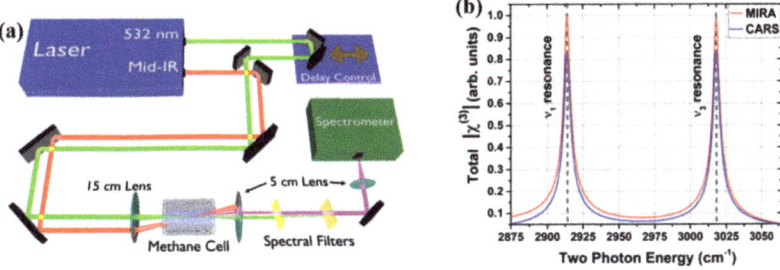

**Figure 11.** (a) The optical path of CARS nonlinear optical system based on MIRA. (b) $\chi^{(3)}$ as a detuning function of MIRA and CARS [56]. Republished with permission from Ref. [56]. Copyright 2022 American Physical Society.

As the main energy substance of the human body, glucose is often used in solvent medicines or energy supplements and plays a significant role in the synthesis and catabolism of cells. The metabolism of glucose itself is closely related to the metabolic decomposition in various pathological reactions, and it is often used in the early detection technology of cancer. Long et al. proposed a two-color stimulated Raman scattering (SRS) imaging method to detect glucose uptake and metabolic activity in mouse brain tissue [57]. As a nonlinear process in optical inelastic scattering, SRS enables high-resolution imaging of probe substrates. By editing $^{13}$C, they constructed a $^{13}$C-labeled 3-O-propargyl-D-glucose (3-OPG-$^{13}$C$_3$) probe with a new Raman characteristic peak to ensure the visual monitoring of glucose uptake by the original 3-OPC probe in vivo. Moreover, the probe can achieve spectral resolution for other types of glucose (D7-glucose).

Figure 12a is their flow chart of experiment. Inspired by the strategy of isotope editing in vibrational spectroscopy, they successfully synthesized 3-OPG-$^{13}$C$_3$ with total productivity is 33%, and the 3-OPG-$^{13}$C$_3$ probe does not have any crosstalk with the spectrum of D7-glucose. This new combinatorial probe enables the study of glucose uptake activity and synergy in adipogenesis by SRS imaging. Figure 12b shows the Raman spectra of different kinds of probes in prostate cancer cell-3 (PC-3) and PBS dispersion, respectively. Among them, the red chromatographic line is double-labeled with D7-glucose and 3-OPG-$^{13}$C$_3$ probes in PC-3 solution, and two different peaks are generated at 2053 cm$^{-1}$ and 2133 cm$^{-1}$, which represent the uptake and metabolism of glucose, respectively. The Raman characteristic peaks of 3-OPG-$^{13}$C$_3$ reflect the uptake of glucose in PC-3, while the characteristic peaks of D7-glucose reflect the metabolism of glucose in PC-3. Edited combinatorial probes are used for simultaneously studying glucose uptake and incorporation activity using SRS microscopy. This research provides a practical experimental idea for the diagnosis and treatment of cancer.

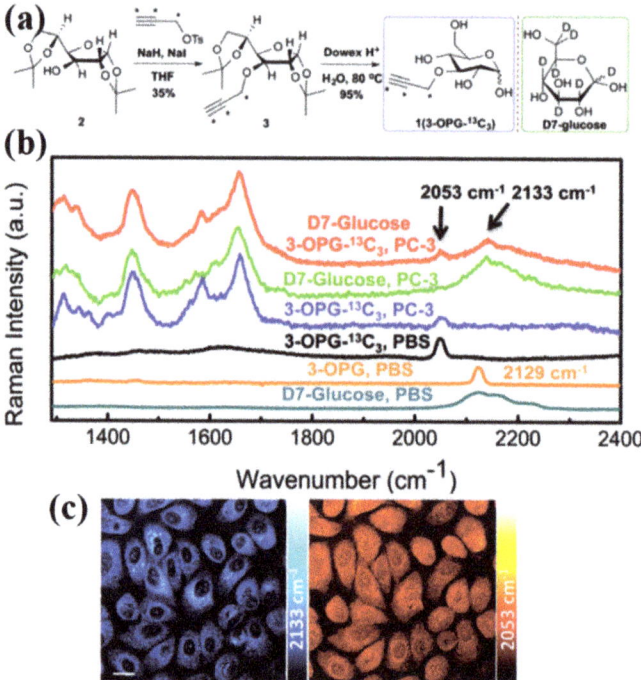

**Figure 12.** (a) The editing and synthesis process of 3-OPG-$^{13}$C$_3$ and the structure of D7-glucose. (b) Raman spectra of different types of glucose probes in PC-3 and PBS solutions. (c) Two-color SRS imaging of PC-3 cells using 3-OPG-$^{13}$C$_3$ and D7-glucose probes at 2133 cm$^{-1}$ and 2053 cm$^{-1}$, respectively [57]. Republished with permission from Ref. [57]. Copyright 2022 Royal Society of Chemistry.

To achieve high-precision characterization to break through the diffraction limit is one of the research directions of micro and nano microscopic systems when studying substances below the optical diffraction limit. Neranga et al. achieved TPEF imaging of single and nanoclusters (NC) glutathione-protected Au$_{25}$ (Au$_{25}$SG$_{18}$) by near-field scanning optical microscopy [58]. They adjusted the pH in the Au$_{25}$SG$_{18}$ solution to induce a change in the charge environment in the solution, which facilitated the separation of single Au$_{25}$SG$_{18}$ NCs, as shown in Figure 13a. This is due to the increased emission density caused by the aggregation of Au$_{25}$ NCs at pH = 5.0. Broader emission spectrum compared to pH = 7.0 solution as shown in Figure 13b. To further verify the effect of solution concentration on the TPEF imaging of NSOM, they selected 1.4 nm and 12 nm solutions for TPEF imaging control, which is shown in Figure 13c,d. The nature of the different imaging results is related to the size of the spacing between the nanoclusters. The imaging results in Figure 13c demonstrate that TPEF for single NCs can be achieved by NSOM when the nanocluster spacing is less than 50 nm. Taking advantage of the unique optical properties of Au$_{25}$, the nanoenvironment can be assessed by non-optical microscopy of TPEF for local ultrasensitive detection. Moreover, a single Au$_{25}$SG$_{18}$ NC can be used in a biosensing system and image detection of the local nanoenvironment in the organism, thereby realizing the early diagnosis of disease.

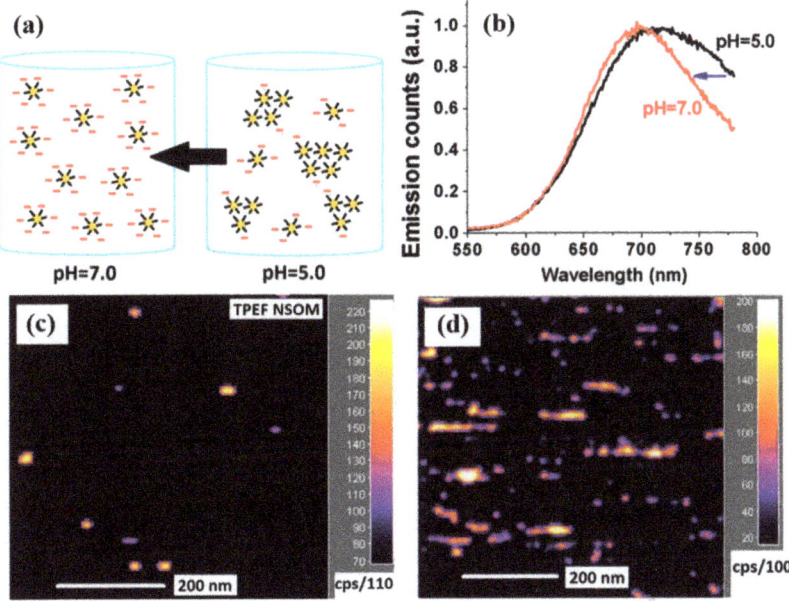

**Figure 13.** (**a**) Changes in pH lead to changes in the spacing of Au$_{25}$ nanoclusters. (**b**) The increase in pH narrows the change in the full width at half the maximum of the emission peak of the molecule. (**c**,**d**) are the TPEF imaging results of 1.4 nm and 12 nm solutions coated on the glass substrate surface, respectively [58]. Republished with permission from Ref. [58]. Copyright 2022 American Chemical Society.

ZnSe doped with a single crystal structure is a popular choice for the preparation of light-emitting diodes, and its optical properties exhibit nonlinear response characteristics with the change of incident light intensity. Irradiation of single-crystal ZnSe in the near-infrared band (~770 nm) with a femtosecond laser results in a two-photon absorption process followed by blue high-energy fluorescence photons (~475 nm). However, single-crystal ZnSe is not easy to obtain, and the preparation cost is higher than that of polycrystalline structure. Geoff's research group studied the two-photon absorption process of polycrystalline ZnSe, and considered whether polycrystalline ZnSe could induce stimulated emission through two-photon absorption at 775 nm under ultra-high light intensity [59]. Due to the short fluorescence lifetime, a 0.5 mm thick ZnSe thin layer sample was mounted in an optical cavity with a length of 10 cm. As illustrated in Figure 14a, a femtosecond laser was used to irradiate a polycrystalline ZnSe material with high-intensity incident light with a wavelength of 775 nm, and the two-photon-induced fluorescence effect was successfully achieved. Figure 14b–e shows that the researchers found that the fluorescence radiation depth of polycrystalline ZnSe gradually increased under different incident light intensities. Under long pulses, as the peak intensity of the incident light intensity diminishes, the fluorescence intensity also gradually decreases and penetrates deeper into the material. Figure 14f shows the lifetime fluorescence intensity versus wavelength decay with time through the nonlinear optical microscopy system. It could be found that the fluorescence lifetime decays gradually with the increase in the pumping intensity of the incident light, which indicates that energy exchange is taking place between the ions. These findings demonstrate that polycrystalline ZnSe could reach the lasing threshold by two-photon pumping at 775 nm using a femtosecond laser under appropriate pumping intensity and crystal cooling conditions. This research has great commercialization potential in developing fluorescent display and lighting equipment.

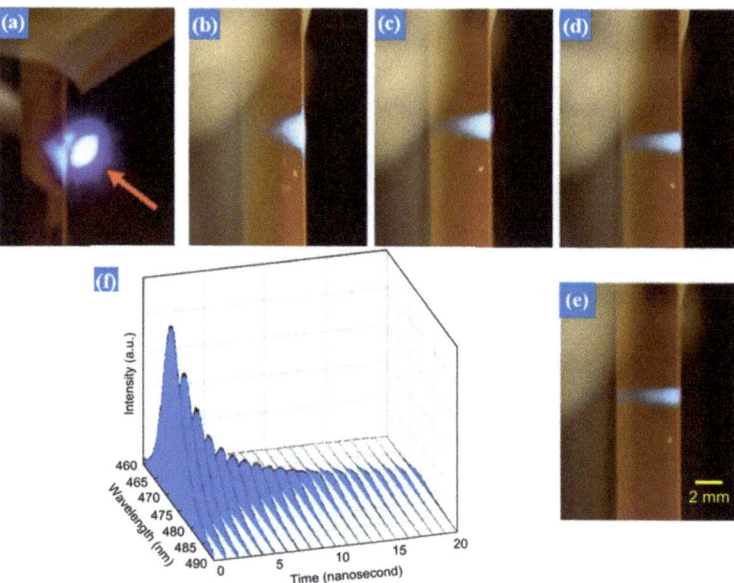

**Figure 14.** The two-photon-induced fluorescence of polycrystalline ZnSe at 775 nm is dependent on the incident laser intensity. (**a**) Blue fluorescence front view of polycrystalline ZnSe. (**b–e**) are the side views of the fluorescence radiation at incident light intensities of 25GW cm$^{-2}$, 4GW cm$^{-2}$, 1.5GW cm$^{-2}$, and 1GW cm$^{-2}$, respectively. As the incident intensity decreases, the depth of fluorescence radiation in the sample becomes larger. (**f**) is the time-dependent decay curve of polycrystalline ZnSe with fluorescence intensity and wavelength [59]. Republished with permission from Ref. [59]. Copyright 2022Chinese Academy of Sciences.

## 4. Summary

The nonlinear optical effect will make the imaging focal spot smaller than the focal spot of the excitation light, thus breaking the classical diffraction limit, greatly improving the spatial resolution of confocal microscopy, and enabling super-resolution three-dimensional imaging. In particular, the surface plasmon of metal nanoparticles based on the frequency doubling effect can amplify nonlinear optical signals, improve the efficiency of nonlinear processes, and enhance the imaging resolution of nonlinear optical microscopes. This review analyses and summarizes the mechanism of plasmon enhancement of nonlinear optical signals by introducing and commenting on related research work. In addition, the applications and prospects of nonlinear optical microscopy systems in biology, medicine, and materials research are reviewed in this paper. These studies and applications attempted in practical scenarios will profoundly impact the update and development of nonlinear optical microscopy systems in the future.

## 5. Outlook

The electromagnetic field enhancement and hot electron transfer in surface plasmons can enhance nonlinear optical effects, especially the combination of some surface plasmon coupling structures and semiconductor materials can significantly improve the nonlinear conversion efficiency. Although a large number of research and theoretical explorations are currently striving to build a complete system for enhancing the conversion efficiency of nonlinear optics, there are still some physical mechanisms that are still being explored and debated. We believe that the system for nonlinear optical enhancement mechanisms will become more and more mature shortly, which is also crucial for the construction of nonlinear optoelectronic devices.

**Funding:** This work was supported by the Foundation of Liaoning Key Laboratory of Chemical Additive Synthesis and Separation (ZJNK2110), the National Science Foundation of China (grant nos. 91436102 and 11374353), the Fundamental Research Funds for the Central Universities (06500067).

**Institutional Review Board Statement:** Not applicable.

**Informed Consent Statement:** Not applicable.

**Data Availability Statement:** The data presented in this study are available on request from the corresponding author.

**Conflicts of Interest:** The authors declare no conflict of interest.

## References

1. Kang, D.; Li, R.; Cao, S.; Sun, M. Nonlinear optical microscopies: Physical principle and applications. *Appl. Spectrosc. Rev.* **2021**, *56*, 52–66. [CrossRef]
2. Ahmadivand, A.; Gerislioglu, B. Deep-and vacuum-ultraviolet metaphotonic light sources. *Mater. Today* **2021**, *51*, 208–221. [CrossRef]
3. Shen, W.; Chen, J.; Wu, J.; Li, X.; Zeng, H. Nonlinear optics in lead halide perovskites: Mechanisms and applications. *ACS Photonics* **2020**, *8*, 113–124. [CrossRef]
4. Autere, A.; Jussila, H.; Dai, Y.; Wang, Y.; Lipsanen, H.; Sun, Z. Nonlinear optics with 2D layered materials. *Adv. Mater.* **2018**, *30*, 1705963. [CrossRef]
5. Schawlow, A.L.; Townes, C.H. Infrared and optical masers. *Phys. Rev.* **1958**, *112*, 1940. [CrossRef]
6. Kleinman, D. Nonlinear dielectric polarization in optical media. *Phys. Rev.* **1962**, *126*, 1977. [CrossRef]
7. Franken, P.; Ward, J. Optical harmonics and nonlinear phenomena. *Rev. Mod. Phys.* **1963**, *35*, 23. [CrossRef]
8. Heiman, D.; Hellwarth, R.; Levenson, M.; Martin, G. Raman-induced Kerr effect. *Phys. Rev. Lett.* **1976**, *36*, 189. [CrossRef]
9. Hellwarth, R. Third-order optical susceptibilities of liquids and solids. *Prog. Quantum Electron.* **1977**, *5*, 1–68. [CrossRef]
10. Shen, Y. Recent advances in nonlinear optics. *Rev. Mod. Phys.* **1976**, *48*, 1. [CrossRef]
11. Parodi, V.; Jacchetti, E.; Osellame, R.; Cerullo, G.; Polli, D.; Raimondi, M.T. Nonlinear optical microscopy: From fundamentals to applications in live bioimaging. *Front. Bioeng. Biotechnol.* **2020**, *8*, 585363. [CrossRef] [PubMed]
12. Li, R.; Wang, X.; Zhou, Y.; Zong, H.; Chen, M.; Sun, M. Advances in nonlinear optical microscopy for biophotonics. *J. Nanophotonics* **2018**, *12*, 033007. [CrossRef]
13. Kauranen, M.; Zayats, A.V. Nonlinear plasmonics. *Nat. Photonics* **2012**, *6*, 737–748. [CrossRef]
14. Dudley, J.M. Light, lasers, and the Nobel Prize. *Adv. Photonics* **2020**, *2*, 050501. [CrossRef]
15. Zhao, J.; Xue, S.; Ji, R.; Li, B.; Li, J. Localized surface plasmon resonance for enhanced electrocatalysis. *Chem. Soc. Rev.* **2021**, *50*, 12070–12097. [CrossRef]
16. Prabowo, B.A.; Purwidyantri, A.; Liu, K.-C. Surface plasmon resonance optical sensor: A review on light source technology. *Biosensors* **2018**, *8*, 80. [CrossRef]
17. Zhao, X.; Hao, Q.; Ni, Z.-H.; Qiu, T. Single-molecule surface-enhanced Raman spectroscopy (SM-SERS): Characteristics and analysis. *Acta Phys. Sin.* **2021**, *70*, 137401. [CrossRef]
18. Li, C.; Le, V.; Wang, X.; Hao, X.; Liu, X.; Kuang, C. Resolution Enhancement and Background Suppression in Optical Super-Resolution Imaging for Biological Applications. *Laser Photonics Rev.* **2021**, *15*, 1900084. [CrossRef]
19. Zhang, C.; Hugonin, J.-P.; Greffet, J.-J.; Sauvan, C. Surface plasmon polaritons emission with nanopatch antennas: Enhancement by means of mode hybridization. *ACS Photonics* **2019**, *6*, 2788–2796. [CrossRef]
20. Kuo, C.-F.; Chu, S.-C. Dynamic control of the interference pattern of surface plasmon polaritons and its application to particle manipulation. *Opt. Express* **2018**, *26*, 19123–19136. [CrossRef]
21. Butet, J.; Brevet, P.-F.; Martin, O.J. Optical second harmonic generation in plasmonic nanostructures: From fundamental principles to advanced applications. *ACS Nano* **2015**, *9*, 10545–10562. [CrossRef] [PubMed]
22. Ahmadivand, A.; Semmlinger, M.; Dong, L.; Gerislioglu, B.; Nordlander, P.; Halas, N.J. Toroidal dipole-enhanced third harmonic generation of deep ultraviolet light using plasmonic meta-atoms. *Nano Lett.* **2018**, *19*, 605–611. [CrossRef] [PubMed]
23. Azzouz, A.; Hejji, L.; Kim, K.-H.; Kukkar, D.; Souhail, B.; Bhardwaj, N.; Brown, R.J.; Zhang, W. Advances in surface plasmon resonance–based biosensor technologies for cancer biomarker detection. *Biosens. Bioelectron.* **2022**, *197*, 113767. [CrossRef] [PubMed]
24. Wang, Q.; Ren, Z.-H.; Zhao, W.-M.; Wang, L.; Yan, X.; Zhu, A.-S.; Qiu, F.-M.; Zhang, K.-K. Research advances on surface plasmon resonance biosensors. *Nanoscale* **2022**, *14*, 564–591. [CrossRef] [PubMed]
25. Zong, C.; Xu, M.; Xu, L.-J.; Wei, T.; Ma, X.; Zheng, X.-S.; Hu, R.; Ren, B. Surface-enhanced Raman spectroscopy for bioanalysis: Reliability and challenges. *Chem. Rev.* **2018**, *118*, 4946–4980. [CrossRef]
26. Willets, K.A.; Van Duyne, R.P. Localized surface plasmon resonance spectroscopy and sensing. *Annu. Rev. Phys. Chem.* **2007**, *58*, 267–297. [CrossRef]
27. Ma, S.; Yang, D.-J.; Ding, S.-J.; Liu, J.; Wang, W.; Wu, Z.-Y.; Liu, X.-D.; Zhou, L.; Wang, Q.-Q. Tunable size dependence of quantum plasmon of charged gold nanoparticles. *Phys. Rev. Lett.* **2021**, *126*, 173902. [CrossRef] [PubMed]

28. Wang, X.; Liu, C.; Gao, C.; Yao, K.; Masouleh, S.S.M.; Berté, R.; Ren, H.; Menezes, L.d.S.; Cortés, E.; Bicket, I.C. Self-constructed multiple plasmonic hotspots on an individual fractal to amplify broadband hot electron generation. *ACS Nano* **2021**, *15*, 10553–10564. [CrossRef]
29. Shi, J.; Guo, Q.; Shi, Z.; Zhang, S.; Xu, H. Nonlinear nanophotonics based on surface plasmon polaritons. *Appl. Phys. Lett.* **2021**, *119*, 130501. [CrossRef]
30. Törmä, P.; Barnes, W.L. Strong coupling between surface plasmon polaritons and emitters: A review. *Rep. Prog. Phys.* **2014**, *78*, 013901. [CrossRef]
31. Joseph, S.; Sarkar, S.; Joseph, J. Grating-Coupled Surface Plasmon-Polariton Sensing at a Flat Metal–Analyte Interface in a Hybrid-Configuration. *ACS Appl. Mater. Interfaces* **2020**, *12*, 46519–46529. [CrossRef] [PubMed]
32. You, J.; Bongu, S.; Bao, Q.; Panoiu, N. Nonlinear optical properties and applications of 2D materials: Theoretical and experimental aspects. *Nanophotonics* **2019**, *8*, 63–97. [CrossRef]
33. Dorfman, K.E.; Schlawin, F.; Mukamel, S. Nonlinear optical signals and spectroscopy with quantum light. *Rev. Mod. Phys.* **2016**, *88*, 045008. [CrossRef]
34. Edappadikkunnummal, S.; Nherakkayyil, S.N.; Kuttippurath, V.; Chalil, D.M.; Desai, N.R.; Keloth, C. Surface plasmon assisted enhancement in the nonlinear optical properties of phenothiazine by gold nanoparticle. *J. Phys. Chem. C* **2017**, *121*, 26976–26986. [CrossRef]
35. Prabu, S.; David, E.; Viswanathan, T.; Jinisha, J.A.; Malik, R.; Maiyelvaganan, K.R.; Prakash, M.; Palanisami, N. Ferrocene conjugated donor-π-acceptor malononitrile dimer: Synthesis, theoretical calculations, electrochemical, optical and nonlinear optical studies. *J. Mol. Struct.* **2020**, *1202*, 127302. [CrossRef]
36. Hao, Q.; Peng, Z.; Wang, J.; Fan, X.; Li, G.; Zhao, X.; Ma, L.; Qiu, T.; Schmidt, O.G. Verification and Analysis of Single-Molecule SERS Events via Polarization-Selective Raman Measurement. *Anal. Chem.* **2022**, *94*, 1046–1051. [CrossRef]
37. Dong, J.; Cao, Y.; Han, Q.; Gao, W.; Li, T.; Qi, J. Nanoscale flexible Ag grating/AuNPs self-assembly hybrid for ultra-sensitive sensors. *Nanotechnology* **2021**, *32*, 155603. [CrossRef]
38. Dong, J.; Yang, C.; Wu, H.; Wang, Q.; Cao, Y.; Han, Q.; Gao, W.; Wang, Y.; Qi, J.; Sun, M. Two-Dimensional Self-Assembly of Au@Ag Core–Shell Nanocubes with Different Permutations for Ultrasensitive SERS Measurements. *ACS Omega* **2022**, *7*, 3312–3323. [CrossRef]
39. Zhao, X.; Dong, J.; Cao, E.; Han, Q.; Gao, W.; Wang, Y.; Qi, J.; Sun, M. Plasmon-exciton coupling by hybrids between graphene and gold nanorods vertical array for sensor. *Appl. Mater. Today* **2019**, *14*, 166–174. [CrossRef]
40. Dong, J.; Wang, Y.; Wang, Q.; Cao, Y.; Han, Q.; Gao, W.; Wang, Y.; Qi, J.; Sun, M. Nanoscale engineering of ring-mounted nanostructure around AAO nanopores for highly sensitive and reliable SERS substrates. *Nanotechnology* **2022**, *33*, 135501. [CrossRef]
41. Ding, S.-J.; Zhang, H.; Yang, D.-J.; Qiu, Y.-H.; Nan, F.; Yang, Z.-J.; Wang, J.; Wang, Q.-Q.; Lin, H.-Q. Magnetic plasmon-enhanced second-harmonic generation on colloidal gold nanocups. *Nano Lett.* **2019**, *19*, 2005–2011. [CrossRef] [PubMed]
42. Mi, X.; Wang, Y.; Li, R.; Sun, M.; Zhang, Z.; Zheng, H. Multiple surface plasmon resonances enhanced nonlinear optical microscopy. *Nanophotonics* **2019**, *8*, 487–493. [CrossRef]
43. Cui, L.; Li, R.; Mu, T.; Wang, J.; Zhang, W.; Sun, M. In situ Plasmon-Enhanced CARS and TPEF for Gram staining identification of non-fluorescent bacteria. *Spectrochim. Acta A Mol. Biomol. Spectrosc.* **2022**, *264*, 120283. [CrossRef] [PubMed]
44. Shen, S.; Shan, J.; Shih, T.-M.; Han, J.; Ma, Z.; Zhao, F.; Yang, F.; Zhou, Y.; Yang, Z. Competitive Effects of Surface Plasmon Resonances and Interband Transitions on Plasmon-Enhanced Second-Harmonic Generation at Near-Ultraviolet Frequencies. *Phys. Rev. Appl.* **2020**, *13*, 024045. [CrossRef]
45. Cao, Y.; Sun, M. Tip-enhanced Raman spectroscopy. *Rev. Phys.* **2022**, *8*, 100067. [CrossRef]
46. Cao, Y.; Cheng, Y.; Sun, M. Graphene-based SERS for sensor and catalysis. *Appl. Spectrosc. Rev.* **2022**, *56*. [CrossRef]
47. Jiang, T.; Kravtsov, V.; Tokman, M.; Belyanin, A.; Raschke, M.B. Ultrafast coherent nonlinear nanooptics and nanoimaging of graphene. *Nat. Nanotechnol.* **2019**, *14*, 838–843. [CrossRef]
48. Wang, Y.; Sun, M.; Meng, L. Tip-enhanced two-photon-excited fluorescence of monolayer $MoS_2$. *Appl. Surf. Sci.* **2022**, *576*, 115835. [CrossRef]
49. Reshef, O.; De Leon, I.; Alam, M.Z.; Boyd, R.W. Nonlinear optical effects in epsilon-near-zero media. *Nat. Rev. Mater.* **2019**, *4*, 535–551. [CrossRef]
50. Psilodimitrakopoulos, S.; Mouchliadis, L.; Paradisanos, I.; Kourmoulakis, G.; Lemonis, A.; Kioseoglou, G.; Stratakis, E. Twist angle mapping in layered WS2 by polarization-resolved second harmonic generation. *Sci. Rep.* **2019**, *9*, 14285. [CrossRef]
51. Psilodimitrakopoulos, S.; Mouchliadis, L.; Maragkakis, G.M.; Kourmoulakis, G.; Lemonis, A.; Kioseoglou, G.; Stratakis, E. Real-time spatially resolved determination of twist angle in transition metal dichalcogenide heterobilayers. *2D Materials* **2020**, *8*, 015015. [CrossRef]
52. Li, R.; Wang, L.; Mu, X.; Chen, M.; Sun, M. Nonlinear optical characterization of porous carbon materials by CARS, SHG and TPEF. *Spectrochim Acta A Mol. Biomol. Spectrosc.* **2019**, *214*, 58–66. [CrossRef] [PubMed]
53. Li, R.; Zhang, Y.; Xu, X.; Zhou, Y.; Chen, M.; Sun, M. Optical characterizations of two-dimensional materials using nonlinear optical microscopies of CARS, TPEF, and SHG. *Nanophotonics* **2018**, *7*, 873–881. [CrossRef]

54. Masia, F.; Pope, I.; Watson, P.; Langbein, W.; Borri, P. Bessel-Beam Hyperspectral CARS Microscopy with Sparse Sampling: Enabling High-Content High-Throughput Label-Free Quantitative Chemical Imaging. *Anal. Chem.* **2018**, *90*, 3775–3785. [CrossRef] [PubMed]
55. Li, R.; Wang, L.; Mu, X.; Chen, M.; Sun, M. Biological nascent evolution of snail bone and collagen revealed by nonlinear optical microscopy. *J. Biophotonics* **2019**, *12*, e201900119. [CrossRef] [PubMed]
56. Traverso, A.J.; Hokr, B.; Yi, Z.; Yuan, L.; Yamaguchi, S.; Scully, M.O.; Yakovlev, V.V. Two-Photon Infrared Resonance Can Enhance Coherent Raman Scattering. *Phys. Rev. Lett.* **2018**, *120*, 063602. [CrossRef]
57. Long, R.; Zhang, L.; Shi, L.; Shen, Y.; Hu, F.; Zeng, C.; Min, W. Two-color vibrational imaging of glucose metabolism using stimulated Raman scattering. *Chem. Commun.* **2018**, *54*, 152–155. [CrossRef]
58. Abeyasinghe, N.; Kumar, S.; Sun, K.; Mansfield, J.F.; Jin, R.; Goodson, T., 3rd. Enhanced Emission from Single Isolated Gold Quantum Dots Investigated Using Two-Photon-Excited Fluorescence Near-Field Scanning Optical Microscopy. *J. Am. Chem. Soc.* **2016**, *138*, 16299–16307. [CrossRef]
59. Li, Q.; Perrie, W.; Li, Z.; Edwardson, S.P.; Dearden, G. Two-photon absorption and stimulated emission in poly-crystalline Zinc Selenide with femtosecond laser excitation. *Opto-Electron. Adv.* **2022**, *5*, 210036. [CrossRef]

Review

# Surface-Plasmon-Assisted Growth, Reshaping and Transformation of Nanomaterials

Chengyun Zhang *, Jianxia Qi, Yangyang Li, Qingyan Han, Wei Gao, Yongkai Wang and Jun Dong *

School of Electronic Engineering, Xi'an University of Posts & Telecommunications, Xi'an 710121, China; superjianxiaqi@163.com (J.Q.); 18066715534@stu.xupt.edu.cn (Y.L.); hanqingyanlove@163.com (Q.H.); gaowei@xupt.edu.cn (W.G.); ykwang@xupt.edu.cn (Y.W.)
* Correspondence: cyzhang@xupt.edu.cn (C.Z.); dongjun@xupt.edu.cn (J.D.)

**Abstract:** Excitation of surface plasmon resonance of metal nanostructures is a promising way to break the limit of optical diffraction and to achieve a great enhancement of the local electromagnetic field by the confinement of optical field at the nanoscale. Meanwhile, the relaxation of collective oscillation of electrons will promote the generation of hot carrier and localized thermal effects. The enhanced electromagnetic field, hot carriers and localized thermal effects play an important role in spectral enhancement, biomedicine and catalysis of chemical reactions. In this review, we focus on surface-plasmon-assisted nanomaterial reshaping, growth and transformation. Firstly, the mechanisms of surface-plasmon-modulated chemical reactions are discussed. This is followed by a discussion of recent advances on plasmon-assisted self-reshaping, growth and etching of plasmonic nanostructures. Then, we discuss plasmon-assisted growth/deposition of non-plasmonic nanostructures and transformation of luminescent nanocrystal. Finally, we present our views on the current status and perspectives on the future of the field. We believe that this review will promote the development of surface plasmon in the regulation of nanomaterials.

**Keywords:** surface plasmon resonance; hot electron transfer; thermal field; regulation of nanostructures

## 1. Introduction

The collective oscillation pattern of conduction band electrons in the nanostructures of certain metals (e.g., Au, Ag and Cu) can effectively enhance the absorption efficiency of the nanostructures, known as surface plasmon resonance (SPR) (Figure 1a) [1,2]. The excitation of SPR can form strong interactions with electromagnetic (EM) radiation and electrons of nanostructures, and lead to a significant enhanced EM field near the nanoparticles (NPs), by converging EM radiation to a size smaller than the wavelength of the incident light [3,4]. The high-energy carriers (electron–hole pairs) and local thermal field will generate during the relaxation of enhanced EM field (Figure 1b) [5–8]. Since the mid-1970s, plasmonic physics including surface-plasmon-enhanced spectroscopy (Raman scattering, infrared and fluorescence), sensing and waveguiding has been widely studied [9–11]. In recent years, the research related to surface plasmon has been extended from plasmonic physics to plasmon-assisted chemical reactions [12–16]. With the enhanced EM field, hot carriers and thermal field, SPR can efficiently modulate surface reactions, especially to enable chemical reactions that are difficult or even impossible to occur under conventional conditions, forming an important frontier in SPR-assisted chemical reactions [17–22].

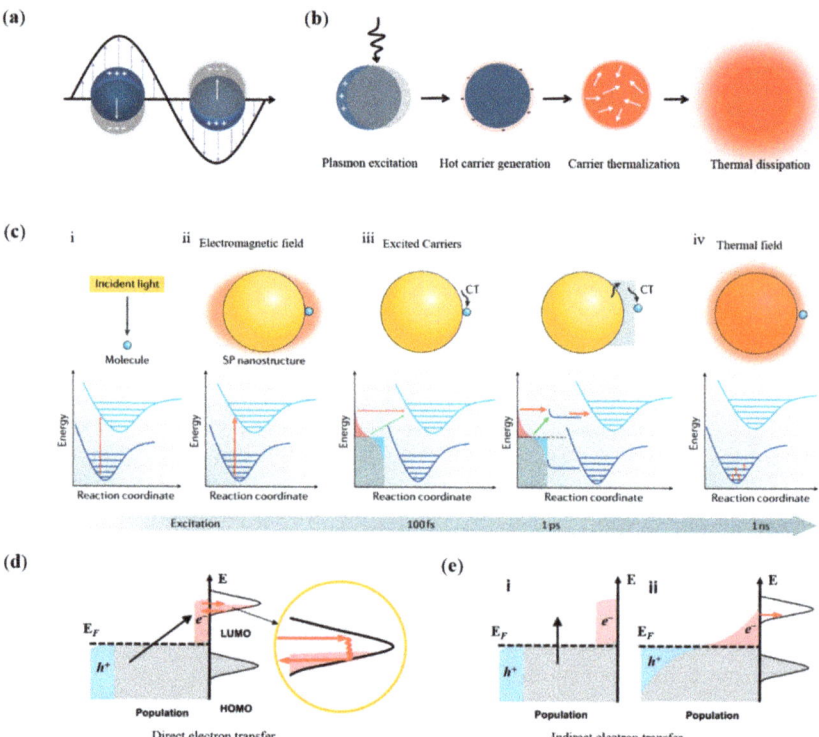

**Figure 1.** (**a**) Coherent oscillation of electrons and the enhanced EM field [23]; (**b**) Scheme illustration of SPR excitation and relaxation [5]; (**c**) Catalytic effects of photon (i), plasmonic EM field (ii), hot carrier transfers (iii) and plamsonic thermal field (iv) [18]. Direct (**d**) and indirect hot carriers transfer (**e**) between metal nanostructure and reactants adsorbed on the surface [1].

High spatial confinement and ultrafast time scale properties of SPR provide a new approach for the precise control of nanomaterial growth and phase transitions. Although plasmonic catalysis on molecules has been extensively studied, systematic reports on the plasmon catalytic effects of nanomaterials are also needed. In this report, we focus on the plasmon catalysis effect on nanomaterials. First, we discuss the mechanisms of plasmon catalysis, including plasmonic EM field, hot carrier transfer and local thermal field. Subsequently, we discuss recent research advances on plasmon catalysis on inorganic nanomaterials, including the reshaping and self-growth/etching of plasmonic nanomaterials, the growth and transformation of non-plasmonic materials. Finally, we present our view on the current state of the field and development in the field of plasmonic-assisted nanomaterial optimization.

## 2. Mechanisms of Surface-Plasmon-Modulated Chemical Reaction

By photoexcitation of SPR, molecules present in solutions or gas streams or on the surface of metallic NPs can be catalyzed to undergo chemical reactions leading to the deposition of metal atoms or other products, resulting in the growth of nanomaterials. In this process, it is still the reacting molecules that are catalyzed by plasmonic field, hot carriers or thermal field, and the reaction mechanism that steams from a plasmon catalyzed reaction of molecules is applicable in the plasmon-assisted growth of nanomaterials. Therefore, we begin with the discussion of surface plasmon excitation and relaxation, and the mechanism of plasmon catalysis with enhanced EM field, hot carrier transfers and localized thermal field (Figure 1c) [18].

## 2.1. The Catalytic Effect of Plasmonic EM Field

For conventional photocatalysis, photon energy excites the electronic transition of the reactants from the ground state to the excited state, and force the molecules to move along the potential energy surface of the excited state (Figure 1c). Finally, molecules in the excited state can either undergo direct chemical reactions or decay back to the ground state with additional vibrational energy, facilitating the overcoming of activation potential [23]. Similar reaction channels of conventional optical excitation are suitable for plasmonic EM field catalyzed reactions [24–27]. SPR excitation brings about a change in the spatial distribution of the optical field, and eventually the photon density on the surface of metal nanostructures will increase substantially [28,29]. Therefore, the local EM field with higher photon densities will lead to a significantly enhanced reaction efficiency of molecules in the vicinity of nanostructures. For catalytic reactions driven by photon or plasmonic EM fields, resonant energy transfer is required by the energy coupling between the surface plasmon and the reactant molecules.

## 2.2. The Catalytic Effect of Hot Carriers

SPR in nanostructures can be radiatively relaxed by re-emission of photons or non-radiatively relax by landau damping or chemical interface damping [30–32]. If the unoccupied electronic orbitals of the reactance that adsorbed at the interface are coupled to the nanostructure, a new electronic state associated with the chemical bonding at the interface will generate [33–35]. The induced electronic states can participate in the coherent oscillations of SPR and accelerate its relaxation, an effect known as interfacial damping [30,36]. During SPR relaxation with the involvement of interfacial damping, hot electrons are generated directly in the orbitals of the electronic unoccupied state of the adsorbates, while hot holes are left in the metal nanostructure (direct electron transfer) (Figure 1d) [1,6]. The electron transfer will produce negative ionic states with instantaneous lifetimes of tens of femtoseconds, which is long enough to allow chemical reactions in the excited state or to increase the vibrational energy of the ground state, thus reducing the reaction potential and facilitating chemical reaction [37]. For this direct electron transfer mechanism, the energies of hot carriers need to overlap with the molecule's unoccupied state orbitals for the electron exchange to occur. Taking advantage of this feature, the chemical reaction channel can be enhanced selectively by controlling the hot carriers' energy to achieve a substantial increase in reaction efficiency and selectivity.

The damping effect of SPR due to the energy exchange between the collective coherent oscillation mode of electrons in metals and other particles (e.g., electrons) is called Landau damping [18,38,39]. The energy exchange will accelerate the relaxation of collective coherent oscillations of electrons, and generate hot carriers in metal NPs [40–42]. After a thermalization process, a hot Fermi–Dirac distribution formed, then hot carriers with continuous energy distribution are transferred to the orbitals of the adsorbed molecules (indirect electron transfer) (Figure 1e) [1,43]. Since the hot electrons are distributed continuously around the Fermi energy level, the two-step indirect transfer process has a high transfer efficiency and is the most efficient excitation pathway catalyzed by plasmon [44–47]. The practical efficiency of indirect electron transfer depends on the position of the unoccupied state orbitals of the molecule relative to the Fermi energy level of the metal. The position of the Fermi energy level of the metallic structure is not easily tuned, which means that the energy of thermalized hot electrons is hard to be controlled by the excitation light or the frequency of SPR, thus limiting the ability of plasmon to catalyze selective enhancement of specific chemical reactions.

## 2.3. The Catalytic Effect of Plasmonic Thermal Field

The thermal effect generated by SPR relaxation can be used to heat NPs and surrounding media, and is currently used in photothermal therapy, nanomaterial growth and vapor generation, etc. [48–54]. In addition, SPR can also reduce the reaction potential barrier and modulate the chemical reactions on the surface of metal [17,55–58]. Currently, SPR have been used to promote the room temperature dissociation of water, oxygen and hydrogen, as well as the chemical reactions of small molecules on the surface [59–62].

## 3. Self-Modulatation of Plasmonic Nanostructure
### 3.1. Plasmonic Thermal-Field-Assisted Reshaping of Plasmonic Nanostructure

The homogeneity of the size and shape of plasmonic metal NPs is the important basis for their unique optical features and has been accepted as a general outcome for judging the quality of the synthesized materials. The reduction of the SPR broadening of colloidal metal NPs that are critically influenced by morphological variations has been a key driver for further progress in nanomaterial colloidal synthesis. Andrés and coworkers report a new method to reshape Au nanorods with plasmonic thermal effect to make their SPR spectra as sharp as those of individual Au nanorods [63]. Femtosecond laser with the same wavelength of SPR band centered at 800 nm was chosen as the source to irradiate Au nanorods. It is found that by precisely controlling the irradiation conditions to balance the relationship between the heat transferred to the surrounding environment and the energy deposited onto the nanorod surface, the nanorods could be effectively reaped and its bandwidth could be significantly reduced (Figure 2a–c). The heat dissipation cooling conditions and the density of surfactants such as CTAB on the surface of Au nanorods are strongly correlated with the shape of the treated Au nanorods. This plasmonic-assisted simple, fast and reproducible method can be used for batch processing of Au nanorods, which is of great importance for its application. Lohmüller and coworkers demonstrated that taking advantage of plasmonic thermal effect, Au nanorods can be bent and reshaped to V-shaped nanoantennas (Figure 2d–f) [64]. It is found that Au nanorods need to be freely dispersed in solution, and it will be bent and printed onto a substrate after light irradiation. With the combination of thermal effect and optical force, Au nanorods will be continuously heated and pushed, finally resulting the melting and reaping. This efficient approach holds great potential for the fabrication of V-shaped antennas in which the bending angle and the orientation of antennas can be independently adjusted by tuning the intensity and polarization of irradiated light.

In addition to the self-reshape of plasmonic nanostructure, plasmonic thermal effect can also be used to realize the assembly and welding of the nanostructure. Morphological inhomogeneities can lead to a broadening of the SPR of colloidal metal NPs, thus limiting the feasibility of the plasmon-related technological application. As we can in Figure 2g, with the femtosecond laser irradiation Au NPs are connected into a continuous thin thread [65]. Au NPs with plasmon resonance at 532 nm are chemically scaffolded into chains with the use of cucurbit uril (CB) molecules, resulting in resonant redshift to 745 nm. The rigid gaps (about 0.9 nm) of Au NPs that are glued together lead to plasmonic hotspots in gaps and promote transient inhomogeneous distribution of thermal effects (Figure 2h). Au chains are then irradiated with 805 nm pulse laser, resulting in localized threading and the production of strings of Au NPs. The process of plasmon-assisted assembly of Au NPs can be tracked via the variation of optical resonance. Furthermore, plasmon-assisted self-assembly of Au nanorods is realized by irradiation with low flux femtosecond laser (Figure 2i,j) [66]. It is found that the flux of laser pulse is the most important factor that affects the self-assembly kinetics. For nanorod trimers with a longitudinal SPR wavelength of 800 nm (resonating with illumination laser), the number of trimers and longer oligomers is greatly reduced and the relative number of AuNR dimers is increased after irradiation by a femtosecond pulse with a fluence of 100 $\mu J/cm^2$. If the pulse fluence is greater than 500 $\mu J/cm^2$, the temperature of the interparticle gaps increases greatly, causing local melting of the Au nanorods tips and thus the Au nanorod is welded. Furthermore, plasmon can also be used to assist the spherification of Au nanorods. As we can see in Figure 2k,l, with the femtosecond laser irradiation the Au nanorods in the focal volume of the focusing objective are reshaped, and the reshaping is selective in the aspect ratio and orientation of the Au nanorods, by taking advantage of the narrow longitudinal SPR linewidth and the dipolar optical response of Au nanorods [67]. Additionally, this longitudinal SPR-mediated selective reshaping is employed to achieve five-dimensional optical information recording and readout.

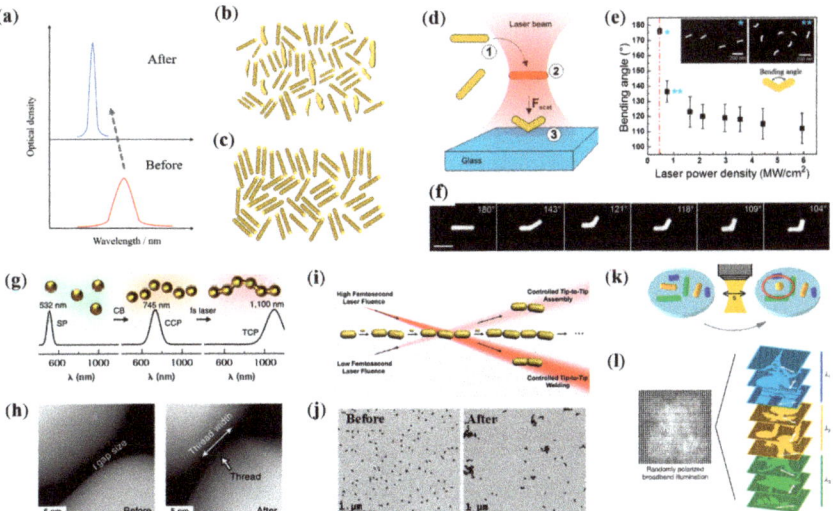

**Figure 2.** (**a**–**c**) Reshaping of Au Nanorods with the irradiation of femtosecond laser. (**a**) Schematic illustration of optical density spectra of Au Nanorod colloids before (red) and after (blue) laser irradiation. (**b**,**c**) Schematic illustration of Au Nanorods morphology before (**b**) and after (**c**) laser irradiation. (**d**–**f**) Plasmon-assisted bending of Au Nanorods [64]. (**d**) Schematic diagram of optical bending of Au Nanorods. (**e**) Au nanorods transition from a straight (*) to a bent (**) morphology and the control of bending angle with the tune of optical power. (**f**) SEM images of bent Au Nanorods. Scale bar is 100 nm. (**g**,**h**) Plasmon-assisted processing of Au NPs strings [65]. (**g**) Schematic illustration. (**h**) TEM images Au NP chains before (left) and after laser irradiation (right). (**i**,**j**) Plasmon-assisted assembling and welding of Au nanorods [66]. (**i**) Schematic illustration. (**j**) TEM images of Au nanorods before and after assembling. (**k**,**l**) Plasmon-assisted spherification of Au nanorods [67]. (**k**) Schematic illustration of plasmon-assisted spherification of Au nanorods. (**l**) Patterns observed on the detector when irradiated by non-polarized broadband light (left) and light with right polarization and wavelength (right).

### 3.2. Hot-Electron-Assisted Self-Growth/Etching of Plasmonic Nanostructure

Hot electrons' generation and transfer during the decay of SPR have attracted much attention recently, which can be applied on plasmonic catalysis, chemical sensing and recording of images based on the oxidation of the NPs [68–70]. SPR mode of an Ag nanocube will be divided into distal and proximal mode (resonance modes localized at the top and bottom of the nanocube), when it is placed on a glass plate at different wavelengths. Tatsuma and coworkers realize a site-selective etching of Ag nanocubes, by taking advantage of plasmon-assisted charge separation and transfer (Figure 3a–c) [71]. Hot-electron-assisted oxidation of Ag nanocubes can be controlled by the location of EM field. Additionally, two SPR modes of Ag nanocubes on $TiO_2$ localized at the top (438 nm) and bottom (648 nm) of the nanocubes can be selective excited by the wavelength of irradiated light. Therefore, the etching of the top or bottom of Ag nanocubes is optional through switching the wavelength of irradiated light. With the assistance of hot electrons and holes generated with SPR decay, Ding and collaborators realized the controllable growth and etching of Au NPs (Figure 3d) [72]. In their study, Au NPs located on Si substrate are immersed in HAuCl4 solution, and 641 nm laser is selected as the irradiation source. When the power of light and the concentration of HAuCl4 is relatively low (4 mW, 5 mM), the size of Au NPs is enlarged. When the light with power the of 6 mW and solution with the concentration 20 mM are used, the Au NPs are etched. Plasmonic hot-electron-assisted selective etching is also observed on single Ag nanoprism (Figure 3e) [73]. To investigate the plasmon-assisted spatially controllable chemical reaction, colloidal Ag nanoprisms which SPR mode can be flexibly adjusted throughout the visible region and near-infrared range are chosen. Meanwhile, redox reactions are selected as a model for the chemical reactions of metal particles, which are driven by the differential electrochemical potential, without additional additives. It is found that there is a direct correlation between the excitation intensity and the shift of the in-plane dipole mode, which provides direct evidence for the role of the SPR mode of optical excitation in influencing the galvanic replacement reaction. The spatial control of replacement reactions within the Ag nanoprism depends on the selectively excitation of the SPR mode, which provides a simple and accurate way to design nanomaterials with unique functionality by switching light of external stimulus.

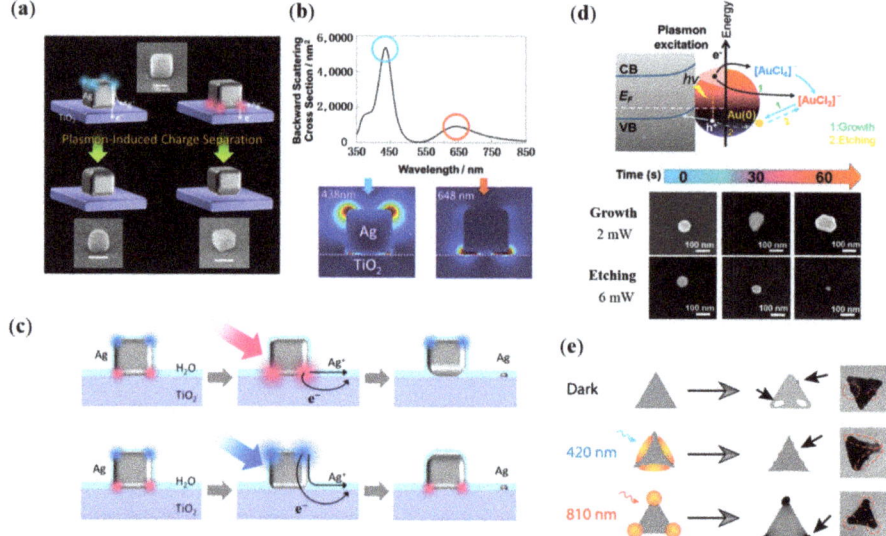

**Figure 3.** Plasmon-assisted etching of plasmonic nanostructure. (**a**–**c**) Plasmon-assisted site-selective etching of Ag nanocubes [71]. (**a**) Schematic diagram. (**b**) Scattering spectra and EM field distribution of the Ag nanocubes on the TiO$_2$. (**c**) Plasmon-assisted electron transition at the interface and site-selective etching of nanocubes. (**d**) Schematic illustration and SEM images of hot carrier-assisted growth/etching of Au NPs [72]. (**e**) Plasmon-assisted spatially controlled replacement reactions on Ag nanoprism [73].

The plasmonic field and catalysis of hot carriers can be leveraged to synthesize NPs with special shape, and structure. Mirkin et al. report a plasmon-assisted method for converting spherical Ag NPs into triangular nanoprisms by a series of cycles of silver redox [74]. It is the SPR of seed particles that catalyzes the reduction of Ag$^+$ to Ag$^0$ and induces the conversion to nanoprisms, while Ag$^+$ are sourced from the oxidation decomposition of small seed particles by O$_2$. The converted colloid shows distinctive optical properties that stems from the nanoprism shape of the Ag NPs (Figure 4a). The size of the converted nanoprisms can be modulated by the frequency of irradiated light. In their subsequent work, they demonstrate that the frequency of irradiated light can be further employed to control defect structure of the plasmonic NPs [75]. Structures with higher number of twin boundaries will be generated, when the irradiated light with higher frequency is used. The correlation between structural defects and irradiated light energy originates from the reduction rate of $Ag^+$ that is controlled by the SPR excitation. Chiral inorganic nanostructures with powerful chromophoric activity and self-assembly capabilities have important applications in enantioselective catalysis and optoelectronics. Recently Xu and coworkers propose a plasmon-assisted synthesis method of chiral Au nanostructure, with anisotropy factor up to 0.44 (Figure 4b,c) [76]. They screened a series of chiral ligands and introduced left/right circularly polarized light to the nanoparticle preparation process. By optimizing the wavelength and polarization of the irradiated light, the formation of symmetry breaking on the high index crystal plane is successfully induced and finally obtained chiral nanostructures with homogeneous morphology. This method can be used to achieve the precise synthesis of chiral nanostructures with high optical anisotropy factors. In the works of Sutter and coworkers, surface plasmon is employed to assisted the growth of plasmonic anisotropic nanostructures (Figure 4d,e) [77,78]. With the help of liquid cell electron microscopy, the metal nanoparticle transformation process in solutions containing seed particles is investigated in situ. With the oxidation of the sparsely dispersed Ag seeds, $Ag^0$ is uniformly attached to the sides of the nanoprism, making it larger in size. The rate of attachment can be modulated by the intensity of the EM field, as the oxidation reaction of Ag seeds is catalyzed by plasmonic hot carriers. They further investigate the influence of reducing agent and non-aggregation agent to the growth process of Ag seeds. A new growth mechanism is explored whereby aggregation of small Ag NPs in dense suspensions promotes the formation of anisotropic nanoprisms, which coexist and compete with the previously observed direct transformation. The report of plasmon-assisted anisotropic growth of Au nanoprisms demonstrate that the plasmon-assisted growth of Ag NPs is also applicable in the synthesis of other plasmonic metals (Figure 4f,g) [79]. Meanwhile, this work established a fundamental mechanism at the molecular level for the plasmon-assisted self-growth of plasmonic nanostructures.

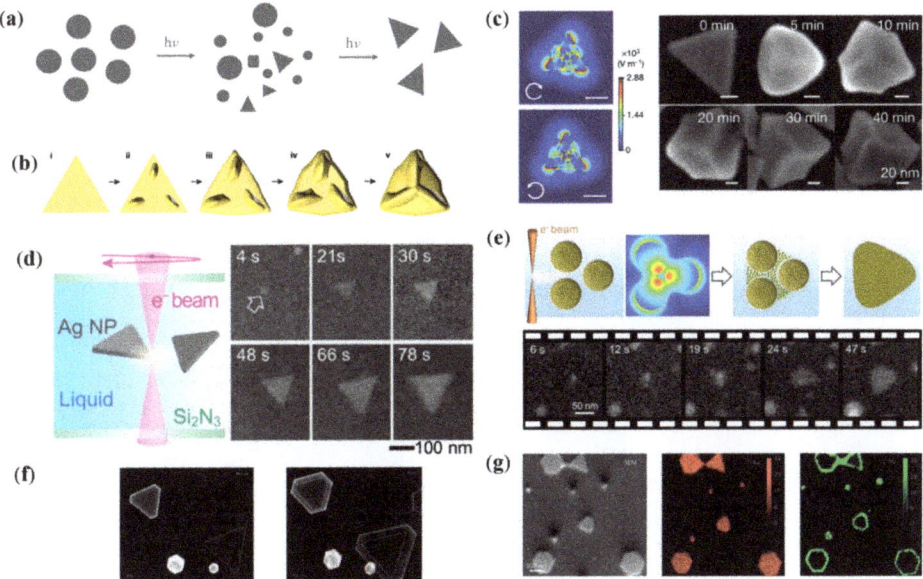

**Figure 4.** Plasmon-assisted growth of plasmonic nanostructures. (**a**) Schematic illustration of plasmon-assisted conversion of Ag NPs to nanoprisms. (**b**,**c**) Plasmon-assisted synthesis of chiral nanostructure.

(**b**) Schematic diagram of chiralization process of Au NPs [76]. (**c**) Calculated EM field of Au NPs under excitation of circularly polarized light and SEM images of Au NPs after laser irradiation for different times. (**d**,**e**) Schematic illustrations and TEM images of plasmon-assisted synthetic Ag nanoprisms [77,78]. (**f**) Morphology of Au nanoprisms before and after plasmon-assisted growth. (**g**) Elemental distribution of Au nanoprisms after plasmon-assisted growth [79].

## 4. Plasmon-Assisted Selective Deposition of Non-Plasmonic Nanomaterials

In addition to the self-reshaping/growth/etching of plasmonic nanostructures, SPR can be further used to induce the selective deposition of non-plasmonic nanostructures. Plasmonic thermal field with the controllable nanoscale heating capability has great potential to control chemical reactions. Temperatures required for deposition can be easily generated with a tightly focused CW laser beam irradiation on the plasmonic structure. Brongersma, Goodwin and their colleagues have proposed a novel chemical vapor deposition method using the efficient plasmonic local heating strategy and applied it to achieve selective deposition of many types of nanomaterials, such as semiconductor nanowires, Si nanotubes, PbO nanowires and $TiO_2$ polycrystalline dots (Figure 5a,b) [80,81]. In this method, a substrate with plasmonic nanomaterials is exposed to a gaseous environment that contains reactant precursor. Then, with the resonance light irradiation, non-plasmonic nanomaterial growth can be observed in the illuminated NPs. The high spatial control and energy-efficient properties of the plasmon-assisted chemical vapor deposition method inspires new pathways for the building of novel photothermal devices. With Au NPs as photothermal sources, a plasmonic thermal-field-assisted solvothermal synthesis is proposed, that can be used to achieve spatially controlled deposition of a wide range of materials, as long as solvothermal synthesis exists (Figure 5c) [82]. Similarly, Sasaki and coworkers proposed a plasmon thermal-field-assisted hydrothermal synthesis, and realized the targeted location of ZnO on the surface of Au nanoantennas, with the help of a confined heat production function at the nanoscale (Figure 5d) [83].

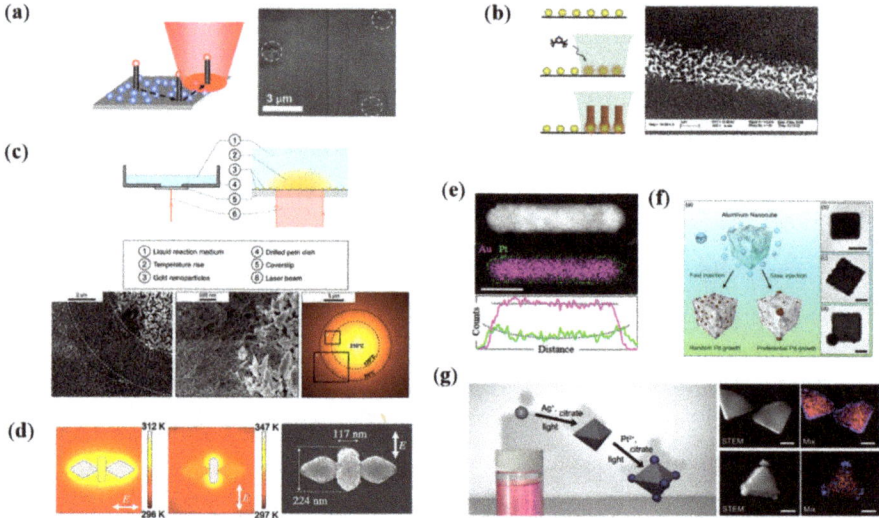

**Figure 5.** Plasmon-assisted selective deposition of non-plasmonic nanomaterials. (**a**,**b**) Plasmonic thermal-field-assisted selectively chemical vapor deposition. (**a**) Deposition of semiconductor nanowires/Si nanotubes [80] (**b**) Deposition of PbO/$TiO_2$ on glass substrates [81]. (**c**) Plasmonic

thermal-field-assisted solvothermal synthesis [82]. Schematic illustration (upper panel) and SEM images of synthetic crystal and the corresponding temperature distribution (lower panel). (**d**) Plasmon-assisted growth of ZnO on Au nanoantennas, calculated temperature distributions (left) and SEM image (right) [83]. (**e–g**) Plasmon-assisted selective deposition of non-plasmonic materials. (**e**) Deposition of Pt. on the ends of Au nanorods, STEM and EDS map of Pt-Au core shell nanorod [84]. (**f**) Schematic diagram and TEM images of selective dipositive Pd on Al nanocubes [85]. (**g**) Deposition of Pt. on Ag cores. Schematic diagram (left) and TEM, EDS maps of products (right) [86].

Then, we discuss plasmon-assisted location of non-plasmonic nanomaterials on the tips of plasmonic structures with the more concentrated EM fields. With the longitudinal SPR excitation, plasmonic hot electrons as the redox agent catalyzed the nucleation of $Pt^0$ in the presence of chloroplatinic acid (Figure 5e) [84]. The anisotropic spatial distribution of hot electrons promotes the selective deposition of platinum onto Au nanorods. In the work of Halas and coworkers, monodisperse Al nanocubes with sharp corners are obtained. Taking advantage of plasmon catalysis of Al nanocrystals, they further realized an Al-Pd heterometallic NPs, by regioselective deposing of Pd nanoclusters on the vertices of nanocubes (Figure 5f) [85]. Similar strategy is used by Personick et al., and the reduction of Pt. ions onto Ag cores is realized (Figure 5g) [86]. Ag-Pt core−shell and core−satellite structures are successfully synthesized by controlling the Pt. ion reduction rate with the excitation of SPR.

## 5. Plasmon-Assisted Transformation of Luminescent Nanocrystal

Rare-earth ion-doped inorganic luminescent nanomaterials are widely noticed and applied because of their rich linear energy level structure, full spectral coverage of fluorescence emission function in the UV-Vis-IR region and stable physicochemical properties [87,88]. In general, matrix materials with single crystal structures are required for high selectivity and polarizable photoemission. For micro/nano systems, subsequent annealing treatment is generally used to achieve the optimization of the crystal structure and improvement of physicochemical stability and luminescence properties. However, the conventional annealing process tends to cause the agglomeration and uncontrollable morphology of particles, and it is hard to realize selective optimization of the particles. Technical bottleneck of conventional method is an urgent problem in the field of nanocrystal growth and luminescent materials research.

Recently, our group has developed a new method for nanocrystal optimization and achieved rapid and controllable nanocrystal transformation by the thermal effect of surface plasmon and the catalytic effect of plasmonic hot electrons [89–92]. Polycrystalline rare-earth-doped fluoride NPs prepared at room temperature were chosen as the main target for the study (Figure 6a). The process and mechanism of the transformation of rare-earth crystals driven by surface plasmon were investigated by using the luminescence of trivalent rare-earth ions as the probe of structure and in situ characterization techniques [89]. The results show that the transformation of polycrystalline $NaYF_4$ to single-crystalline $Y_2O_3$ NPs can be achieved under the irradiation of visible light at the milliwatt level and within milliseconds thanks to the efficient light absorption of Au NPs (Figure 6b,c) [89]. The optimization of crystal structure and crystallinity of the products leads to a significant increase in fluorescence radiation intensity, red–green ratio and monochromaticity (Figure 6d). Then, the effects of resonant and non-resonant excitation of surface plasmon on the crystal transformation and the synergistic mechanism of interband absorption of Au NPs on the crystal transformation are investigated. The results show that the thermal effects and hot electrons generated during the interband transition of Au NPs are lower in energy compared with the resonant excitation process of SPR, resulting in significantly lower crystal transformation rates and poor crystallinity of products under 442 nm laser irradiation (Figure 6e). The kinetic processes of crystal structure change are also investigated. The results show that the thermal effect and hot electrons generated during the relaxation of SPR act synergistically to promote the crystal particle transformation. The thermal effect of the surface plasmon drives the optimization of the crystal structure and induces the transformation from polycrystalline to single crystal (Figure 6f,g) [90]. Meanwhile, the hot electrons catalyze the oxidation process of fluoride to oxide by interacting with oxygen molecules to generate $O_2^-$ ions with strong oxidation properties.

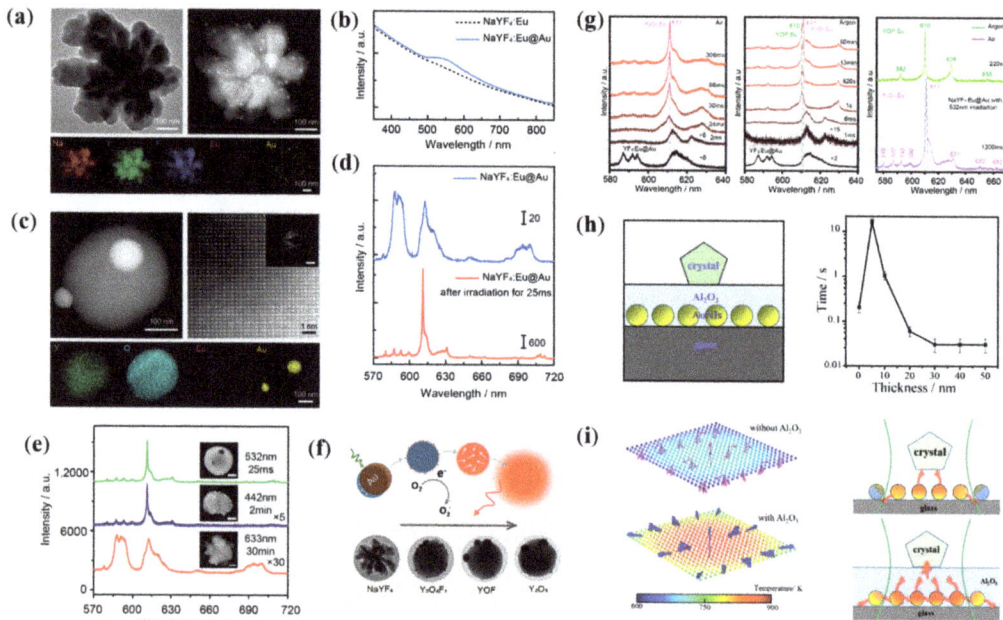

**Figure 6.** Plasmon-assisted crystal transformation from polycrystalline fluoride to single crystal oxide. (**a,b**) TEM, EDX maps and absorption spectra of NaYF$_4$:Eu and NaYF$_4$:Eu@Au; (**c**) STEM, EDX and SAED maps of transformed single-crystalline Y$_2$O$_3$ NP; (**d**) Fluorescence spectra of NaYF$_4$:Eu and NaYF$_4$:Eu@Au and transformed Y$_2$O$_3$:Eu. (**e**) Fluorescence spectra and SEM images of transformed products after light irradiation with different wavelengths of light. (**f**) Schematic diagram of plasmon-assisted polycrystalline fluoride NPs transformation. (**g**) Evolution of luminescence spectra during the plasmon-assisted crystal (YF$_3$:Eu@Au) transformation in air (left) and argon (middle), and luminescence spectra of YF$_3$:Eu particle after annealed at 600 °C for 1 h (right) [90]. (**h**) Schematic diagram of heat trapping structure (left) and the required irradiation time for the crystal transformation with different thickness of dielectric layer Al$_2$O$_3$ (right). (**i**) Simulated thermal transfer modes (left) and schematic diagram of the mechanism (right) of pure Au nanoislands and Au nanoislands/Al$_2$O$_3$ systems, the direction of the arrow representing the direction of heat flow [91].

To enhance the utilization of plasmonic thermal energy, we designed a heat trapping structure by introducing a dielectric layer Al$_2$O$_3$ to the surface of self-assembled Au nanoislands (Figure 6h,i) [91]. The absorption cross section of light gradually increases with the thickening of the Al$_2$O$_3$ layer, and the heat is mainly transferred to the dielectric layer Al$_2$O$_3$ due to the increase in the effective refractive index. In addition, the Al$_2$O$_3$ layer transfers a large amount of heat generated by the more Au NPs around the laser spot to the crystal. Both the enhanced light absorption and heat utilization result in higher temperatures and faster crystal transformation rates. When the thickness of Al$_2$O$_3$ is too thin (5–10 nm), the heat diffuses through the Al$_2$O$_3$ layer into the surrounding environment, resulting in greater heat loss and slower crystal transition efficiency. Overall, these results suggest that a heat-trapping structure with a sufficiently thick Al$_2$O$_3$ can ensure effective light absorption and heat utilization, thus improving the efficiency of the SPR-assisted photothermal conversion. For the transformation of luminescent nanomaterials, the direct catalytic target of surface plasmon is inorganic nanocrystals rather than molecules, which opens up a new field of plasmon-assisted catalysis.

## 6. Conclusions

In this review, we give an overview of surface plasmon excitation and relaxation, and the mechanism of plasmon catalysis with enhanced EM field, direct/indirect hot carrier transfers and localized thermal field. Meanwhile, we have introduced recent advances of plasmon-assisted modulation of inorganic nanomaterials including the plasmonic nanomaterials and other non-plasmonic nanomaterials. Especially, the application of surface plasmon in the rapid in situ achievement of luminescent single crystal NPs is reviewed in detail.

Related effects of surface plasmon resonance have many potential and irreplaceable advantages in inducing nanocrystal reshape/growth/transitions, and the related experimental and theoretical studies are of great significance for the acquisition of high-quality nanomaterials and the applications of surface plasmon. Although a lot of results have been achieved in the existing works, there are many shortcomings that need to be improved and explored subsequently. For example, although the kinetics of carrier relaxation processes in bulk metals have been extensively studied, these time scales may change in nanomaterials due to confinement effects, and therefore further studies of hot carrier relaxation kinetic processes in nanosystems are needed. Secondly, plasmonic hot carriers have a strong catalytic effect, but the low transfer efficiency, short lifetime of carriers and the energy distribution that cannot be easily regulated have limited the application of carriers in regulating the structure of nanomaterials. Moreover, plasmonic metal nanostructures have strong photothermal conversion efficiency, but it is difficult to achieve a large and stable temperature gradient on the metal surface due to the fast diffusion of heat from the high thermal conductivity of the metal. Thus, it limits the spatial selectivity of plasmonic thermal field catalysis. We believe that once the hot carriers and local temperature distribution are effectively controlled, the field of plasmonic catalysis will achieve great success in assisting the highly tunable and selective growth and transformation of nanomaterials.

**Author Contributions:** C.Z. wrote and prepared the manuscript for submission; J.Q. and Y.L. were involved in planning the manuscript. Q.H., W.G. and Y.W. contributed to the discussion of the contents. J.D. reviewed and supervised the manuscript. All authors have read and agreed to the published version of the manuscript.

**Funding:** This work was founded by the National Natural Science Foundation of China (Grant 12104366 and 12004303), Shaanxi Provincial Natural Science Foundation (Grant 2022JQ-041 and 2022JZ-05), the Shaanxi provincial research plan for young scientific and technological new stars (2021KJXX-45), the Young Talent fund of University Association for Science and Technology in Shaanxi (Grant 20200511), Key R&D Program of Shaanxi Province (Grant 2022SF-333).

**Acknowledgments:** We thank Professor Hairong Zheng (Shaanxi Normal University), and Zhenglong Zhang (Shaanxi Normal University) for valuable discussions and suggestions.

**Conflicts of Interest:** The authors declare no conflict of interest.

## References

1. Zhang, Y.; He, S.; Guo, W.; Hu, Y.; Huang, J.; Mulcahy, J.R.; Wei, W.D. Surface-Plasmon-Driven Hot Electron Photochemistry. *Chem. Rev.* **2017**, *118*, 2927–2954. [CrossRef]
2. Misewich, J.A.; Heinz, T.F.; Newns, D.M. Desorption induced by multiple electronic transitions. *Phys. Rev. Lett.* **1992**, *68*, 3737–3740. [CrossRef]
3. Dong, J.; Cao, Y.; Han, Q.; Wang, Y.; Qi, M.; Zhang, W.; Qiao, L.; Qi, J.; Gao, W. Plasmon-exciton coupling for nanophotonic sensing on chip. *Opt. Express* **2020**, *28*, 20817–20829. [CrossRef] [PubMed]
4. Zhang, Z.; Fang, Y.; Wang, W.; Chen, L.; Sun, M. Propagating Surface Plasmon Polaritons: Towards Applications for Remote-Excitation Surface Catalytic Reactions. *Adv. Sci.* **2016**, *3*, 1500215. [CrossRef] [PubMed]
5. Brongersma, M.L.; Halas, N.J.; Nordlander, P. Plasmon-induced hot carrier science and technology. *Nat. Nanotechnol.* **2015**, *10*, 25–34. [CrossRef] [PubMed]

6. Zhou, D.; Li, X.; Zhou, Q.; Zhu, H. Infrared driven hot electron generation and transfer from non-noble metal plasmonic nanocrystals. *Nat. Commun.* **2020**, *11*, 2944. [CrossRef]
7. Govorov, A.O.; Richardson, H.H. Generating heat with metal nanoparticles. *Nano Today* **2007**, *2*, 30–38. [CrossRef]
8. Baffou, G.; Quidant, R.; Javier Garcia de Abajo, F. Nanoscale Control of Optical Heating in Complex Plasmonic Systems. *ACS Nano* **2010**, *4*, 709–716. [CrossRef]
9. Marcos, M.; Alvarez, J.T.K.; Schaaff, T.G.; Marat, N.; Shafigullin, I.V.; Robert, L. Whetten Optical Absorption Spectra of Nanocrystal Gold Molecules. *J. Phys. Chem. B* **1997**, *101*, 3706–3712.
10. Ding, S.; Yi, J.; Li, J.; Ren, B.; Wu, D.; Panneerselvam, R.; Tian, Z. Nanostructure-based plasmon-enhanced Raman spectroscopy for surface analysis of materials. *Nat. Rev. Mater.* **2016**, *1*, 16021. [CrossRef]
11. Dong, J.; Zhang, Z.; Zheng, H.; Sun, M. Recent Progress on Plasmon-Enhanced Fluorescence. *Nanophotonics* **2015**, *4*, 472–490. [CrossRef]
12. Huang, Y.F.; Zhang, M.; Zhao, L.B.; Feng, J.M.; Wu, D.Y.; Ren, B.; Tian, Z.Q. Activation of oxygen on gold and silver nanoparticles assisted by surface plasmon resonances. *Angew. Chem. Int. Ed. Engl.* **2014**, *53*, 2353–2357. [CrossRef] [PubMed]
13. Aslam, U.; Chavez, S.; Linic, S. Controlling energy flow in multimetallic nanostructures for plasmonic catalysis. *Nat. Nanotechnol.* **2017**, *12*, 1000–1005. [CrossRef]
14. Kale, M.J.; Avanesian, T.; Christopher, P. Direct Photocatalysis by Plasmonic Nanostructures. *ACS Catal.* **2013**, *4*, 116–128. [CrossRef]
15. Shirhatti, P.R.; Rahinov, I.; Golibrzuch, K.; Werdecker, J.; Geweke, J.; Altschaffel, J.; Kumar, S.; Auerbach, D.J.; Bartels, C.; Wodtke, A.M. Observation of the adsorption and desorption of vibrationally excited molecules on a metal surface. *Nat. Chem.* **2018**, *10*, 592–598. [CrossRef]
16. Xie, W.; Schlucker, S. Hot electron-induced reduction of small molecules on photorecycling metal surfaces. *Nat. Commun.* **2015**, *6*, 7570. [CrossRef] [PubMed]
17. Zhang, X.; Li, X.; Reish, M.E.; Zhang, D.; Su, N.Q.; Gutierrez, Y.; Moreno, F.; Yang, W.; Everitt, H.O.; Liu, J. Plasmon-Enhanced Catalysis: Distinguishing Thermal and Nonthermal Effects. *Nano Lett.* **2018**, *18*, 1714–1723. [CrossRef]
18. Zhan, C.; Chen, X.-J.; Yi, J.; Li, J.-F.; Wu, D.-Y.; Tian, Z.-Q. From plasmon-enhanced molecular spectroscopy to plasmon-mediated chemical reactions. *Nat. Rev. Chem.* **2018**, *2*, 216–230. [CrossRef]
19. Wang, S.; Ding, T. Photothermal-Assisted Optical Stretching of Gold Nanoparticles. *ACS Nano* **2019**, *13*, 32–37. [CrossRef]
20. Novo, C.; Funston, A.M.; Mulvaney, P. Direct observation of chemical reactions on single gold nanocrystals using surface plasmon spectroscopy. *Nat. Nanotechnol.* **2008**, *3*, 598–602. [CrossRef]
21. Linic, S.; Christopher, P.; Ingram, D.B. Plasmonic-metal nanostructures for efficient conversion of solar to chemical energy. *Nat. Mater.* **2011**, *10*, 911–921. [CrossRef] [PubMed]
22. Mukherjee, S.; Libisch, F.; Large, N.; Neumann, O.; Brown, L.V.; Cheng, J.; Lassiter, J.B.; Carter, E.A.; Nordlander, P.; Halas, N.J. Hot electrons do the impossible: Plasmon-induced dissociation of $H_2$ on Au. *Nano Lett.* **2013**, *13*, 240–247. [CrossRef] [PubMed]
23. Aslam, U.; Rao, V.G.; Chavez, S.; Linic, S. Catalytic conversion of solar to chemical energy on plasmonic metal nanostructures. *Nat. Catal.* **2018**, *1*, 656–665. [CrossRef]
24. Koichi, A.; Makoto, F.; Carsten, R.; Junji, T.; Hirotaka, M.; Yoshimichi, O.; Naoya, Y.; Toshiya, W. A plasmonic photocatalyst consisting of silver nanoparticles embedded in titanium dioxide. *J. Am. Chem. Soc.* **2008**, *130*, 1676–1680.
25. Liu, Z.W.; Hou, W.B.; Pavaskar, P.; Aykol, M.; Cronin, S.B. Plasmon Resonant Enhancement of Photocatalytic Water Splitting Under Visible Illumination. *Nano Lett.* **2011**, *11*, 1111–1116. [CrossRef]
26. Tesema, T.E.; Kafle, B.; Tadesse, M.G.; Habteyes, T.G. Plasmon-Enhanced Resonant Excitation and Demethylation of Methylene Blue. *J. Phys. Chem. C* **2017**, *121*, 7421–7428. [CrossRef]
27. Kazuma, E.; Jung, J.; Ueba, H.; Trenary, M.; Kim, Y. Direct Pathway to Molecular Photodissociation on Metal Surfaces Using Visible Light. *J. Am. Chem. Soc.* **2017**, *139*, 3115–3121. [CrossRef]
28. Zhang, Z.; Deckert-Gaudig, T.; Deckert, V. Label-free monitoring of plasmonic catalysis on the nanoscale. *Analyst* **2015**, *140*, 4325–4335. [CrossRef]
29. Zhang, Z.; Richard-Lacroix, M.; Deckert, V. Plasmon induced polymerization using a TERS approach: A platform for nanostructured 2D/1D material production. *Faraday Discuss.* **2017**, *205*, 213–226. [CrossRef]
30. Foerster, B.; Joplin, A.; Kaefer, K.; Celiksoy, S.; Link, S.; Sonnichsen, C. Chemical Interface Damping Depends on Electrons Reaching the Surface. *ACS Nano* **2017**, *11*, 2886–2893. [CrossRef]
31. Cortés, E. Activating plasmonic chemistry. *Science* **2018**, *362*, 28–29. [CrossRef] [PubMed]
32. Linic, S.; Chavez, S.; Elias, R. Flow and extraction of energy and charge carriers in hybrid plasmonic nanostructures. *Nat. Mater.* **2021**, *20*, 916–924. [CrossRef] [PubMed]
33. Kang, Y.M.; Najmaei, S.; Liu, Z.; Bao, Y.J.; Wang, Y.M.; Zhu, X.; Halas, N.J.; Nordlander, P.; Ajayan, P.M.; Lou, J.; et al. Plasmonic Hot Electron Induced Structural Phase Transition in a $MoS_2$ Monolayer. *Adv. Mater.* **2014**, *26*, 6467–6471. [CrossRef] [PubMed]
34. Kale, M.J.; Christopher, P. Plasmons at the interface. *Science* **2015**, *349*, 587–588. [CrossRef] [PubMed]
35. Wu, K.; Chen, J.; McBride, J.R.; Lian, T. Efficient hot-electron transfer by a plasmon-induced interfacial charge-transfer transition. *Science* **2015**, *349*, 632–635. [CrossRef]
36. Foerster, B.; Spata, V.A.; Carter, E.A.; Sönnichsen, C.; Link, S.J.S.A. Plasmon damping depends on the chemical nature of the nanoparticle interface. *Sci. Adv.* **2019**, *5*, eaav0704. [CrossRef]

37. Chen, Y.C.; Hsu, Y.K.; Popescu, R.; Gerthsen, D.; Lin, Y.G.; Feldmann, C. Au@Nb@H x K1-xNbO$_3$ nanopeapods with near-infrared active plasmonic hot-electron injection for water splitting. *Nat. Commun.* **2018**, *9*, 232. [CrossRef]
38. Wei, Q.; Wu, S.; Sun, Y. Quantum-Sized Metal Catalysts for Hot-Electron-Driven Chemical Transformation. *Adv. Mater.* **2018**, *30*, e1802082. [CrossRef]
39. Kim, Y.; Smith, J.G.; Jain, P.K. Harvesting multiple electron-hole pairs generated through plasmonic excitation of Au nanoparticles. *Nat. Chem.* **2018**, *10*, 763–769. [CrossRef]
40. Manjavacas, A.; Liu, J.G.; Kulkarni, V.; Nordlander, P. Plasmon-Induced Hot Carriers in Metallic Nanoparticles. *ACS Nano* **2014**, *8*, 7630–7638. [CrossRef]
41. Tagliabue, G.; DuChene, J.S.; Abdellah, M.; Habib, A.; Gosztola, D.J.; Hattori, Y.; Cheng, W.H.; Zheng, K.; Canton, S.E.; Sundararaman, R.; et al. Ultrafast hot-hole injection modifies hot-electron dynamics in Au/p-GaN heterostructures. *Nat. Mater.* **2020**, *19*, 1312–1318. [CrossRef] [PubMed]
42. Liu, Y.; Chen, Q.; Cullen, D.A.; Xie, Z.; Lian, T. Efficient Hot Electron Transfer from Small Au Nanoparticles. *Nano Lett.* **2020**, *20*, 4322–4329. [CrossRef] [PubMed]
43. Mukherjee, S.; Zhou, L.; Goodman, A.M.; Large, N.; Ayala-Orozco, C.; Zhang, Y.; Nordlander, P.; Halas, N.J. Hot-electron-induced dissociation of H$_2$ on gold nanoparticles supported on SiO$_2$. *J. Am. Chem. Soc.* **2014**, *136*, 64–67. [CrossRef] [PubMed]
44. Chalabi, H.; Schoen, D.; Brongersma, M.L. Hot-electron photodetection with a plasmonic nanostripe antenna. *Nano Lett.* **2014**, *14*, 1374–1380. [CrossRef]
45. Heilpern, T.; Manjare, M.; Govorov, A.O.; Wiederrecht, G.P.; Gray, S.K.; Harutyunyan, H. Determination of hot carrier energy distributions from inversion of ultrafast pump-probe reflectivity measurements. *Nat. Commun.* **2018**, *9*, 1853. [CrossRef]
46. Clavero, C. Plasmon-induced hot electron generation at nanoparticle/metal-oxide interfaces for photovoltaic and photocatalytic devices. *Nat. Photonics* **2014**, *8*, 95–103. [CrossRef]
47. Rossi, T.P.; Erhart, P.; Kuisma, M. Hot-Carrier Generation in Plasmonic Nanoparticles: The Importance of Atomic Structure. *ACS Nano* **2020**, *14*, 9963–9971. [CrossRef]
48. Takami, A.; Kurita, H.; Koda, S. Laser-Induced Size Reduction of Noble Metal Particles. *J. Phys. Chem. B* **1999**, *103*, 1226–1232. [CrossRef]
49. Boyer, D.; Tamarat, P.; Maali, A.; Lounis, B.; Orrit, M. Photothermal imaging of nanometer-sized metal particles among scatterers. *Science* **2002**, *297*, 1160–1163. [CrossRef]
50. Wang, Z.; Horseman, T.; Straub, A.P.; Yip, N.Y.; Li, D.; Elimelech, M.; Lin, S. Pathways and challenges for efficient solar-thermal desalination. *Sci. Adv.* **2019**, *5*, eaax0763. [CrossRef]
51. Zhou, X.; Zhao, F.; Guo, Y.; Rosenberger, B.; Yu, G. Architecting highly hydratable polymer networks to tune the water state for solar water purification. *Sci. Adv.* **2019**, *5*, eaaw5484. [CrossRef] [PubMed]
52. Zhou, L.; Swearer, D.F.; Zhang, C.; Robatjazi, H.; Zhao, H.; Henderson, L.; Dong, L.; Christopher, P.; Carter, E.A.; Nordlander, P. Quantifying hot carrier and thermal contributions in plasmonic photocatalysis. *Science* **2018**, *362*, 69–72. [CrossRef]
53. Baffou, G.; Quidant, R. Thermo-plasmonics: Using metallic nanostructures as nano-sources of heat. *Laser Photonics Rev.* **2013**, *7*, 171–187. [CrossRef]
54. Neumann, O.; Feronti, C.; Neumann, A.D.; Dong, A.; Schell, K.; Lu, B.; Kim, E.; Quinn, M.; Thompson, S.; Grady, N.; et al. Compact solar autoclave based on steam generation using broadband light-harvesting nanoparticles. *Proc. Natl. Acad. Sci. USA* **2013**, *110*, 11677–11681. [CrossRef] [PubMed]
55. Yu, Y.; Sundaresan, V.; Willets, K.A. Hot Carriers versus Thermal Effects: Resolving the Enhancement Mechanisms for Plasmon-Mediated Photoelectrochemical Reactions. *J. Phys. Chem. C* **2018**, *122*, 5040–5048. [CrossRef]
56. Aibara, I.; Mukai, S.; Hashimoto, S. Plasmonic-Heating-Induced Nanoscale Phase Separation of Free Poly(N-isopropylacrylamide) Molecules. *J. Phys. Chem. C* **2016**, *120*, 17745–17752. [CrossRef]
57. Ou, W.; Zhou, B.; Shen, J.; Lo, T.W.; Lei, D.; Li, S.; Zhong, J.; Li, Y.Y.; Lu, J. Thermal and Nonthermal Effects in Plasmon-Mediated Electrochemistry at Nanostructured Ag Electrodes. *Angew. Chem.-Int. Ed.* **2020**, *59*, 6790–6793. [CrossRef]
58. Zhan, C.; Liu, B.W.; Huang, Y.F.; Hu, S.; Ren, B.; Moskovits, M.; Tian, Z.Q. Disentangling charge carrier from photothermal effects in plasmonic metal nanostructures. *Nat. Commun.* **2019**, *10*, 2671. [CrossRef]
59. Golubev, A.A.; Khlebtsov, B.N.; Rodriguez, R.D.; Chen, Y.; Zahn, D.R.T. Plasmonic Heating Plays a Dominant Role in the Plasmon-Induced Photocatalytic Reduction of 4-Nitrobenzenethiol. *J. Phys. Chem. C* **2018**, *122*, 5657–5663. [CrossRef]
60. Kamarudheen, R.; Castellanos, G.W.; Kamp, L.P.J.; Clercx, H.J.H.; Baldi, A. Quantifying Photothermal and Hot Charge Carrier Effects in Plasmon-Driven Nanoparticle Syntheses. *ACS Nano* **2018**, *12*, 8447–8455. [CrossRef]
61. Chen, X.; Xia, Q.; Cao, Y.; Min, Q.; Zhang, J.; Chen, Z.; Chen, H.Y.; Zhu, J.J. Imaging the transient heat generation of individual nanostructures with a mechanoresponsive polymer. *Nat. Commun.* **2017**, *8*, 1498. [CrossRef] [PubMed]
62. Zhang, X.; Wang, M.; Tang, F.; Zhang, H.; Fu, Y.; Liu, D.; Song, X. Transient Electronic Depletion and Lattice Expansion Induced Ultrafast Bandedge Plasmons. *Adv. Sci.* **2019**, *7*, 1902408. [CrossRef] [PubMed]
63. González-Rubio, G.; Díaz-Núñez, P.; Rivera, A.; Prada, A.; Tardajos, G.; González-Izquierdo, J.; Bañares, L.; Llombart, P.; Macdowell, L.G.; Alcolea, P.M.J.S. Femtosecond laser reshaping yields gold nanorods with ultranarrow surface plasmon resonances. *Science* **2017**, *358*, 640. [CrossRef] [PubMed]
64. Babynina, A.; Fedoruk, M.; Kuehler, P.; Meledin, A.; Doeblinger, M.; Lohmueller, T. Bending Gold Nanorods with Light. *Nano Lett.* **2016**, *16*, 6485–6490. [CrossRef] [PubMed]

65. Herrmann, L.O.; Valev, V.K.; Tserkezis, C.; Barnard, J.S.; Kasera, S.; Scherman, O.A.; Aizpurua, J.; Baumberg, J.J. Threading plasmonic nanoparticle strings with light. *Nat. Commun.* **2014**, *5*, 4568. [CrossRef]
66. Gonzalez-Rubio, G.; Gonzalez-Izquierdo, J.; Banares, L.; Tardajos, G.; Rivera, A.; Altantzis, T.; Bals, S.; Pena-Rodriguez, O.; Guerrero-Martinez, A.; Liz-Marzan, L.M. Femtosecond Laser-Controlled Tip-to-Tip Assembly and Welding of Gold Nanorods. *Nano Lett.* **2015**, *15*, 8282–8288. [CrossRef]
67. Zijlstra, P.; Chon, J.W.; Gu, M. Five-dimensional optical recording mediated by surface plasmons in gold nanorods. *Nature* **2009**, *459*, 410–413. [CrossRef]
68. Zhang, C.; Jia, F.; Li, Z.; Huang, X.; Lu, G. Plasmon-generated hot holes for chemical reactions. *Nano Res.* **2020**, *13*, 3183–3197. [CrossRef]
69. Ma, X.C.; Dai, Y.; Yu, L.; Huang, B.B. Energy transfer in plasmonic photocatalytic composites. *Light Sci. Appl.* **2016**, *5*, e16017. [CrossRef]
70. Kim, Y.; Kazuma, E. Mechanistic studies of plasmon chemistry on metal catalysts. *Angew. Chem. Int. Ed. Engl.* **2018**, *58*, 4800–4808.
71. Saito, K.; Tanabe, I.; Tatsuma, T. Site-Selective Plasmonic Etching of Silver Nanocubes. *J. Phys. Chem. Lett.* **2016**, *7*, 4363–4368. [CrossRef] [PubMed]
72. Long, Y.; Wang, S.; Wang, Y.; Deng, F.; Ding, T. Light-Directed Growth/Etching of Gold Nanoparticles via Plasmonic Hot Carriers. *J. Phys. Chem. C* **2020**, *124*, 19212–19218. [CrossRef]
73. Bhanushali, S.; Mahasivam, S.; Ramanathan, R.; Singh, M.; Harrop Mayes, E.L.; Murdoch, B.J.; Bansal, V.; Sastry, M. Photomodulated Spatially Confined Chemical Reactivity in a Single Silver Nanoprism. *ACS Nano* **2020**, *14*, 11100–11109. [CrossRef]
74. Jin, R.C.; Cao, Y.W.; Mirkin, C.A.; Kelly, K.L.; Schatz, G.C.; Zheng, J.G. Photoinduced conversion of silver nanospheres to nanoprisms. *Science* **2001**, *294*, 1901–1903. [CrossRef] [PubMed]
75. Personick, M.L.; Langille, M.R.; Zhang, J.; Wu, J.; Li, S.; Mirkin, C.A. Plasmon-mediated synthesis of silver cubes with unusual twinning structures using short wavelength excitation. *Small* **2013**, *9*, 1947–1953. [CrossRef] [PubMed]
76. Xu, L.; Wang, X.; Wang, W.; Sun, M.; Choi, W.J.; Kim, J.Y.; Hao, C.; Li, S.; Qu, A.; Lu, M.; et al. Enantiomer-dependent immunological response to chiral nanoparticles. *Nature* **2022**, *601*, 366–373. [CrossRef] [PubMed]
77. Sutter, P.; Li, Y.; Argyropoulos, C.; Sutter, E. In Situ Electron Microscopy of Plasmon-Mediated Nanocrystal Synthesis. *J. Am. Chem. Soc.* **2017**, *139*, 6771–6776. [CrossRef]
78. Sun, M.; Li, Y.; Zhang, B.; Argyropoulos, C.; Sutter, P.; Sutter, E. Plasmonic Effects on the Growth of Ag Nanocrystals in Solution. *Langmuir* **2020**, *36*, 2044–2051. [CrossRef]
79. Zhai, Y.; DuChene, J.S.; Wang, Y.C.; Qiu, J.; Johnston-Peck, A.C.; You, B.; Guo, W.; DiCiaccio, B.; Qian, K.; Zhao, E.W.; et al. Polyvinylpyrrolidone-induced anisotropic growth of gold nanoprisms in plasmon-driven synthesis. *Nat. Mater.* **2016**, *15*, 889–895. [CrossRef]
80. Cao, L.; Barsic, D.N.; Guichard, A.R.; Brongersma, M.L. Plasmon-assisted local temperature control to pattern individual semiconductor nanowires and carbon nanotubes. *Nano Lett.* **2007**, *7*, 3523–3527. [CrossRef]
81. Boyd, D.A.; Greengard, L.; Brongersma, M.; El-Naggar, M.Y.; Goodwin, D.G. Plasmon-Assisted Chemical Vapor Deposition. *Nano Lett.* **2006**, *6*, 2592–2597. [CrossRef] [PubMed]
82. Robert, H.M.L.; Kundrat, F.; Bermudez-Urena, E.; Rigneault, H.; Monneret, S.; Quidant, R.; Polleux, J.; Baffou, G. Light-Assisted Solvothermal Chemistry Using Plasmonic Nanoparticles. *ACS Omega* **2016**, *1*, 2–8. [CrossRef] [PubMed]
83. Fujiwara, H.; Suzuki, T.; Pin, C.; Sasaki, K. Localized ZnO Growth on a Gold Nanoantenna by Plasmon-Assisted Hydrothermal Synthesis. *Nano Lett.* **2020**, *20*, 389–394. [CrossRef] [PubMed]
84. Forcherio, G.T.; Baker, D.R.; Boltersdorf, J.; Leff, A.C.; McClure, J.P.; Grew, K.N.; Lundgren, C.A. Targeted Deposition of Platinum onto Gold Nanorods by Plasmonic Hot Electrons. *J. Phys. Chem. C* **2018**, *122*, 28901–28909. [CrossRef]
85. Robatjazi, H.; Lou, M.; Clark, B.D.; Jacobson, C.R.; Swearer, D.F.; Nordlander, P.; Halas, N.J. Site-Selective Nanoreactor Deposition on Photocatalytic Al Nanocubes. *Nano Lett.* **2020**, *20*, 4550–4557. [CrossRef]
86. Habib, A.; King, M.E.; Etemad, L.L.; Distler, M.E.; Morrissey, K.H.; Personick, M.L. Plasmon-Mediated Synthesis of Hybrid Silver–Platinum Nanostructures. *J. Phys. Chem. C* **2020**, *124*, 6853–6860. [CrossRef]
87. Dong, J.; Gao, W.; Han, Q.; Wang, Y.; Qi, J.; Yan, X.; Sun, M. Plasmon-enhanced upconversion photoluminescence: Mechanism and application. *Rev. Phys.* **2019**, *4*, 100026. [CrossRef]
88. Chen, H.; Sun, M.; Ma, J.; Zhang, B.; Wang, C.; Guo, L.; Ding, T.; Zhang, Z.; Zheng, H.; Xu, H. Multiplasmons-Pumped Excited-State Absorption and Energy Transfer Upconversion of Rare-Earth-Doped Luminescence beyond the Diffraction Limit. *ACS Photonics* **2021**, *8*, 1335–1343. [CrossRef]
89. Zhang, C.; Lu, J.; Jin, N.; Dong, L.; Fu, Z.; Zhang, Z.; Zheng, H. Plasmon-Driven Rapid In Situ Formation of Luminescence Single Crystal Nanoparticle. *Small* **2019**, *15*, 1901286. [CrossRef]
90. Zhang, C.; Kong, T.; Fu, Z.; Zhang, Z.; Zheng, H. Hot electron and thermal effects in plasmonic catalysis of nanocrystal transformation. *Nanoscale* **2020**, *12*, 8768–8774. [CrossRef]
91. Kong, T.; Zhang, C.; Lu, J.; Kang, B.; Fu, Z.; Li, J.; Yan, L.; Zhang, Z.; Zheng, H.; Xu, H. An enhanced plasmonic photothermal effect for crystal transformation by a heat-trapping structure. *Nanoscale* **2021**, *13*, 4585–4591. [CrossRef] [PubMed]
92. Dong, L.; Zhang, C.; Yan, L.; Zhang, B.; Chen, H.; Mi, X.; Fu, Z.; Zhang, Z.; Zheng, H. Quantifying plasmon resonance and interband transition contributions in photocatalysis of gold nanoparticle. *Chin. Phys. B* **2021**, *30*, 077301. [CrossRef]

MDPI
St. Alban-Anlage 66
4052 Basel
Switzerland
www.mdpi.com

*Nanomaterials* Editorial Office
E-mail: nanomaterials@mdpi.com
www.mdpi.com/journal/nanomaterials

Disclaimer/Publisher's Note: The statements, opinions and data contained in all publications are solely those of the individual author(s) and contributor(s) and not of MDPI and/or the editor(s). MDPI and/or the editor(s) disclaim responsibility for any injury to people or property resulting from any ideas, methods, instructions or products referred to in the content.

www.ingramcontent.com/pod-product-compliance
Lightning Source LLC
LaVergne TN
LVHW070600100526
838202LV00012B/527